Being Prepared for Climate Change

A Workbook for Developing Risk-Based Adaptation Plans

United States
Environmental Protection
Agency

Being Prepared for Climate Change

A Workbook for Developing Risk-Based Adaptation Plans

Climate Ready Estuaries
EPA Office of Water

August 2014

TABLE OF CONTENTS

Purpose

Identifying risks associated with climate change and managing them to reduce their impacts is essential. This WORKBOOK presents a guide to climate change adaptation planning based on EPA's experience with watershed management, the National Estuary Program and the Climate Ready Estuaries program. The WORKBOOK will assist organizations that manage environmental resources to prepare a broad, risk-based adaptation plan.

The audience for this WORKBOOK is professionals at organizations that manage environmental resources, especially organizations with a coastal or watershed focus. They are knowledgeable about their systems but not necessarily sophisticated about climate science or risk management. They may be addressing a myriad of issues that require immediate attention and have limited time to focus on adaptation planning for the future. Furthermore, they may need to adapt to climate change impacts within their organization's existing resources. Despite these challenges, managers who realize that climate change will affect their ability to meet their goals will see the need to incorporate climate change risk into their planning.

Although risk management and risk-based vulnerability assessments have been highlighted or recommended by experts in the field of climate change adaptation,[1] to date only a handful of risk-based plans have been published. Interviews with coastal managers conducted by Climate Ready Estuaries staff in 2011 revealed that managers are not sure what is meant by a "risk-based vulnerability assessment," and would like tools to help them proceed.

Uses

The WORKBOOK responds to these impediments as well as to gaps and needs—identified below—that were recognized in climate change strategic planning.

- In the interagency *National Action Plan: Priorities for Managing Freshwater Resources in a Changing Climate* (2011), under Recommendation 3, "Strengthen assessment of vulnerability of water resources to climate change," EPA agreed to lead Action 11: "Continue development of tools and approaches that build capacity for water-related institutions to conduct vulnerability assessments and implement appropriate responses" (page 24).

- EPA's *National Water Program 2012 Strategy: Response to Climate Change* (2012) calls for the Office of Water to "Address climate change adaptation and build stakeholder capacity when

---

[1] Recommendations of risk-based approaches appear in, for example:

Intergovernmental Panel on Climate Change. 2007. *Summary for Policy Makers*.

Government Accountability Office. 2009. *Observations on Federal Efforts to Adapt to a Changing Climate*.

National Research Council. 2010. *Informing an Effective Response to Climate Change*.

National Research Council. 2010. *Adapting to the Impacts of Climate Change*.

Interagency Ocean Policy Task Force. 2010. *National Ocean Policy*.

Interagency Climate Change Adaptation Task Force. 2010. *Progress Report of the Interagency Climate Change Adaptation Task Force: Recommended Actions in Support of a National Climate Change Adaptation Strategy*.

Government Accountability Office. 2013. *Climate Change: Future Federal Adaptation Efforts Could Better Support Local Infrastructure Decision Makers*.

Executive Order 13653. 2013.

implementing [National Estuary Program] Comprehensive Conservation and Management Plans and through the Climate Ready Estuaries Program" (page 49).

- In the *National Fish, Wildlife and Plants Climate Adaptation Strategy* (2012), Goal 4 reads, "Tools, such as vulnerability and risk assessments and scenario planning, can inform and enable management planning and decision-making under uncertainty. Identifying, developing, and employing these types of tools will help managers facilitate adaptation of individual species, increase habitat resilience, and help identify where changes to the built environment may conflict with ecosystem needs" (page 68).

- The *National Ocean Council Implementation Plan* (April 2013) states, "Agencies will take a number of actions to improve the resilience of coastal communities and enhance their ability to adapt to the impacts from climate change, extreme weather events, and ocean acidification....They will offer tools and training courses on how to design and implement vulnerability assessments and develop a national assessment of coastal and ocean vulnerability to both climate change and ocean acidification" (page 16).

- Executive Order 13653 (November 1, 2013) states, "The Federal Government must build on recent progress and pursue new strategies to improve the Nation's preparedness and resilience. In doing so, agencies should promote: (1) engaged and strong partnerships and information sharing at all levels of government; (2) risk-informed decisionmaking and the tools to facilitate it; (3) adaptive learning, in which experiences serve as opportunities to inform and adjust future actions; and (4) preparedness planning."

This WORKBOOK helps meet the need for guidance on conducting climate change vulnerability assessments at a watershed scale, provides decision-support tools, helps people plan climate change adaptation strategies, and builds the capacity of local environmental managers. The WORKBOOK helps EPA to fulfill the commitments that it has made to assist local and regional organizations to effectively plan for climate change impacts.

An organization's own goals and an understanding of the political and ecological context in which it operates are essential pieces of the process. This WORKBOOK helps managers to systematically and progressively develop and build that information into an action plan. When users have gone through the WORKBOOK, they will have produced a planning-level document that can guide their responses to climate change risks. This will help them anticipate change and reach their organizational goals.

Content

The WORKBOOK presents a step-by-step application of a risk management methodology to climate change adaptation. By taking a risk-based approach to assessing vulnerability, users have a formal way to choose among adaptation actions. Selected actions are not simply beneficial, they rise to the top because they will be best for reducing risk.

The WORKBOOK uses the risk management process in an international standard (ISO 31000—*Risk Management*), and tailors it to respond to the practices and norms of the climate change community. Specifically, the ISO methodology (which is conceived as a start-to-finish application of a risk management process) was modified to accommodate a two-part process consisting of (1) a stand-alone vulnerability assessment, followed (either immediately or after a period of time) by (2) an action plan. When faced with many discrete risks, managers can benefit from a process that will help them prioritize those that should receive attention, and decide on how to mitigate them. Each place-based organization's vulnerability assessment and action plan will be unique: the risk management process and this WORKBOOK embrace that philosophy.

The risk management process in the WORKBOOK is informed by a review of existing resources and relevant information. NOAA's *Roadmap for Adapting to Coastal Risk*, NOAA's *Adapting to Climate Change: A Planning Guide for State Coastal Managers*, ICLEI's *Preparing for Climate Change: A Guidebook for Local, Regional, and State Governments*, NWF's *Scanning the Conservation Horizon: A Guide to Climate Change Vulnerability Assessment*, and U.S. Global Change Research Program resources were especially useful.

EPA's National Estuary Program also informs the approach presented in the WORKBOOK. The NEP has four cornerstones: (1) focus on the watershed, (2) integrate science into the decision-making process, (3) foster collaborative problem solving, and (4) involve the public. The WORKBOOK responds to those principles by:

- advancing a risk management approach that is appropriate for the scale of a coastal watershed;
- accommodating the large number of risks likely to be encountered;
- calling for organizations to have scientists and other experts advise on risks to consider and their consequences; and
- promoting a process that is especially suited to public engagement and building consensus.

An organization like a National Estuary Program, whose management structure has robust mechanisms for involving citizens and scientists, may be able to write an adaptation plan using its regular consultation processes.

The development of this WORKBOOK benefitted greatly from ongoing peer input solicited from EPA colleagues, watershed managers and federal interagency partners. Numerous presentations about the WORKBOOK were made to the climate change adaptation community of practice over the past two years. A pilot project to field test the vulnerability assessment methodology ran from September 2012 to September 2013; early results were presented at a stakeholder workshop on vulnerability assessments in February 2013. The vulnerability assessment half of this WORKBOOK was the focus of that workshop, which provided valuable insight into questions that stakeholders would have as they used the WORKBOOK. In October 2013 a draft of the complete WORKBOOK was made available on the EPA website to provide further opportunities for input. Finally, the WORKBOOK draft was submitted to five coastal/watershed management experts for formal independent peer review.

Limitations and Caveats

The WORKBOOK is part of a growing and dynamic body of literature on how to evaluate vulnerability and respond to climate change. Although risk management itself has been successfully used for decades, adaptation to climate change is a rapidly developing field. New material is constantly being published. Many excellent governmental and non-governmental tools and publications are available that explain how to conduct community outreach, identify and comment on the severity of expected climate impacts, or provide instruction on how to assess the vulnerability of a specific species, site or sector to a particular climate change risk. This WORKBOOK identifies other helpful resources and directs users to them.

Similarly, scientific understanding of the magnitude of climate change and its impacts is also growing as we learn more about how global and local environments are responding and how the climate is projected to change. This WORKBOOK points users to information about climate change from the U.S. Global Change Research Program as a primary source. It also draws on other peer-reviewed assessments and government reports when needed. EPA's Climate Ready Estuaries website will link to the latest information from USGCRP and other sources as it is published.

Every place is unique, as is every organization and every situation. The range of technical expertise, resources, ecotypes and stakeholders varies immensely among target audiences of the WORKBOOK. A decision support tool that accounts for this diversity is essential. The WORKBOOK is prescriptive about steps to take, but users have great flexibility to make their needed decisions along the way. The process assumes that users have a good understanding of their place, resource or environment and are familiar with the stressors that are in play and how their system may respond. This WORKBOOK is a roadmap that will help users to (1) apply a risk-based methodology to their organization's goals, (2) weigh various risks, (3) consider associated impacts on goals, and (4) thoughtfully plan for each risk.

Contact Information

For information, questions or comments about this document, contact Michael Craghan, Ph.D., of Climate Ready Estuaries, at U.S. Environmental Protection Agency, Office of Water, Office of Wetlands, Oceans and Watersheds, 1200 Pennsylvania Avenue NW, Mail Code 4504T, Washington, DC 20460, or by email using crehelp@epa.gov.

August 2014

Climate change will bring additional challenges to places and ecosystems that are already under environmental stress. The expected climate changes will worsen existing problems as well as bring new problems. The process described in this WORKBOOK leads you to take a broad look at how climate change will affect your environmental system and your organization. The creation of a planning-level risk-based vulnerability assessment will help you develop an action plan with effective solutions that your stakeholders and partners can help implement.

The likelihood and consequences of all risks are to at least some extent unknowable. However, understanding the suite of climate change risks presents a special type of challenge, one for which most places have few examples or little experience to draw upon:

- The risks will stem from many stressors (changes in air and water temperature, intensification of the hydrologic cycle, sea level rise, and other associated problems) and impacts will be numerous and diverse.

- When risks will materialize and with what intensity are not precisely known. Some risks will have slow, steady onsets (e.g., those driven by sea level rise or warming air temperatures). Other risks will have sudden onsets as thresholds are crossed (e.g., invasive species arrive) or familiar risks will interact with new environmental regimes (e.g., flooding when runoff from stronger precipitation interacts with higher river levels caused by sea level rise).

- We do not know how robust or fragile systems will be when climate stresses begin to accumulate. As well, we do not completely understand how climate stressors will be mediated by the environment or the sensitivity and adaptive capacity of living organisms.

The characteristics of the climate change problem also take away many of the commonly used risk management responses:

- A place-based organization—by itself—will have essentially no effective way to prevent climate change, and some effects will be inescapable.

- Climate risks cannot be studied using repeatable observations the way industrial processes or chemical reactions can. Incredibly diverse factors combine to influence how climate will change and interact to produce climate change's effects: no matter how much time and resources you spend on the problem, it will not be possible to precisely know what will happen 15, 30 or 60 years in the future.

- Since we do not know how intense the change will be in coming decades, and we are not sure how our systems will respond, it is not going to be possible to quantify either costs or benefits in any highly accurate (or easy) way, especially at the spatial scale for which the WORKBOOK is intended.

- Some risks will be very expensive to mitigate; others will be impossible to mitigate. It could take decades to mitigate some risks (e.g., drought impacts on drinking water supply). With some risks, adaptive management techniques will be required.

- Another element that makes managing environmental change different from other risk management contexts is that the public must be involved. Responses must change something in the environment, and in most cases, the public is going to have to accept those changes as well as pay for them. When a private firm manages its own risks it might seek to take every cost-effective action it can. In environmental management, public considerations and the actual cost will make that kind of strategy impossible.

Of course, all environmental planning is plagued by mysteries. Oil spills, hurricanes, population changes, fires or invasive species are equally unpredictable in terms of what, whether or when they will occur, and how bad they might be. Yet plans are made and potential responses are developed for the risks. These risks and hazards are effectively managed by identifying, analyzing, evaluating and mitigating them.[2] Climate change risks can be managed the same way.

Who Should Use This Workbook?

This methodology is appropriate for any type of place-based planning—including hazard mitigation. However, the WORKBOOK was designed with environmental professionals who manage watersheds or coastal places and protect the health of aquatic ecosystems as its main audience. This WORKBOOK assumes that users:

> ### "Organization"
>
> ISO 31000 "can be used by any public, private or community enterprise, association, group or individual" (p. 1). The WORKBOOK, like the standard it is based on, uses the term "organization" to refer to any user.

- represent an organization that has environmental goals or objectives;
- start with some sense that climate change will pose a threat to what their organization is trying to accomplish;
- are comfortable using science to inform decision-making; and
- have sufficient knowledge of their environmental system (or can partner with those who do) to understand how climate changes may affect the way it functions.

The risk management methodology adopted in the WORKBOOK is best used:

- at a spatial scale that is large enough that risks are numerous and diverse and small enough that managers know the territory;
- where using qualitative risk analysis is well suited;
- where many stakeholders are involved; and
- where responses have to be prioritized because not all can be implemented.

EPA and others have resources (see Appendix C) that may be more appropriate than this WORKBOOK when organizations have a narrower scope than a watershed scale planning effort (e.g., focus on a single species or habitat type).

Risk Management

A risk management process helps with decision-making when organizations are faced with uncertainty about whether they will be able to meet their goals. Risk management guides an organization to help it determine what risks are important and need to be addressed. The process encourages a broad look at all potential risks to an organization's goals. The process is systematic and builds on information generated in previous steps, allows for judgments and decisions, and allows users to incorporate new data and information as they become available.

---

[2] FEMA's *Mitigation Planning How-To Guide # 1* (FEMA 386-1), Introduction Table 1, describes a very similar process for hazard mitigation planning.

Risk management is a flexible process. How it is implemented depends on the organization and the purpose for which it is used. Sometimes, when people initially hear that they will be participating in a risk management process, they misperceive what that will mean. Risk management can be a lengthy, demanding process. It often entails laboratory analysis, extensive use of statistical analysis, and other difficult, costly or time-consuming steps. However, when there is no direct, imminent threat to health or safety or a planning-level assessment is acceptable, then less intensive approaches to risk management can be appropriate. The identified risk management framework can be used for chemical exposure, engineering decisions or global climate change, but how each step is executed or what types of conclusions can be reached may be different.

Risk management is about your organization: your goals and objectives, your context, your priorities, your decisions on how to respond. Systematically working through your situation with a risk management framework will help you:

- find risks that you might have overlooked and avoid surprises;
- assess risks differently than you otherwise would have;
- find strategies that can address more than one risk and increase efficiency;
- make better decisions; and
- increase the odds that you will be able to reach your goals.

You will also have a reference to why you think something is important and needs attention as well as a communication tool that allows others to understand system challenges.

The ISO 31000—*Risk Management* methodology is the foundation of this WORKBOOK. Here, the start-to-finish procedure described in the standard is divided into two halves to be consistent with the community of practice for climate change adaptation (Figure I-1). Part 1 is referred to as a vulnerability assessment (which in the WORKBOOK's methodology can be a stand-alone product); Part 2 is an action plan. Part 2 also includes implementation of the action plan.

Risk management

"Risk management allows entities to operate more effectively in environments of uncertainty by providing the discipline and structure in which to address these issues, since risk management is not an end in itself, but an important component of an entity's management process. As such, risk management is interrelated with, among other things, an entity's governance, performance management, and internal controls. The process of risk management provides the rigor and structure necessary to identify and select among alternative risk responses whose cumulative effect is intended to reduce risk, and the methodologies and techniques for making selection decisions. This process enables entities to enhance their capability to identify potential adverse events, assess risks, and establish integrated responses. Further, this phase in the planning process would include support and buy-in from upper levels of management and stakeholders. Acceptance for concepts of the model from this group provides the groundwork for future discussions."

U.S. Government Accountability Office. 2006. *Risk Management: Further Refinements Needed to Assess Risks and Prioritize Protective Measures at Ports and Other Critical Infrastructure.* GAO-06-91, pp. 104–05.

Adaptation to climate change

Part 1: WORKBOOK Steps 1–5
- Vulnerability assessment.

Part 2: WORKBOOK Steps 6–10
- Action plan.
- Implementation.

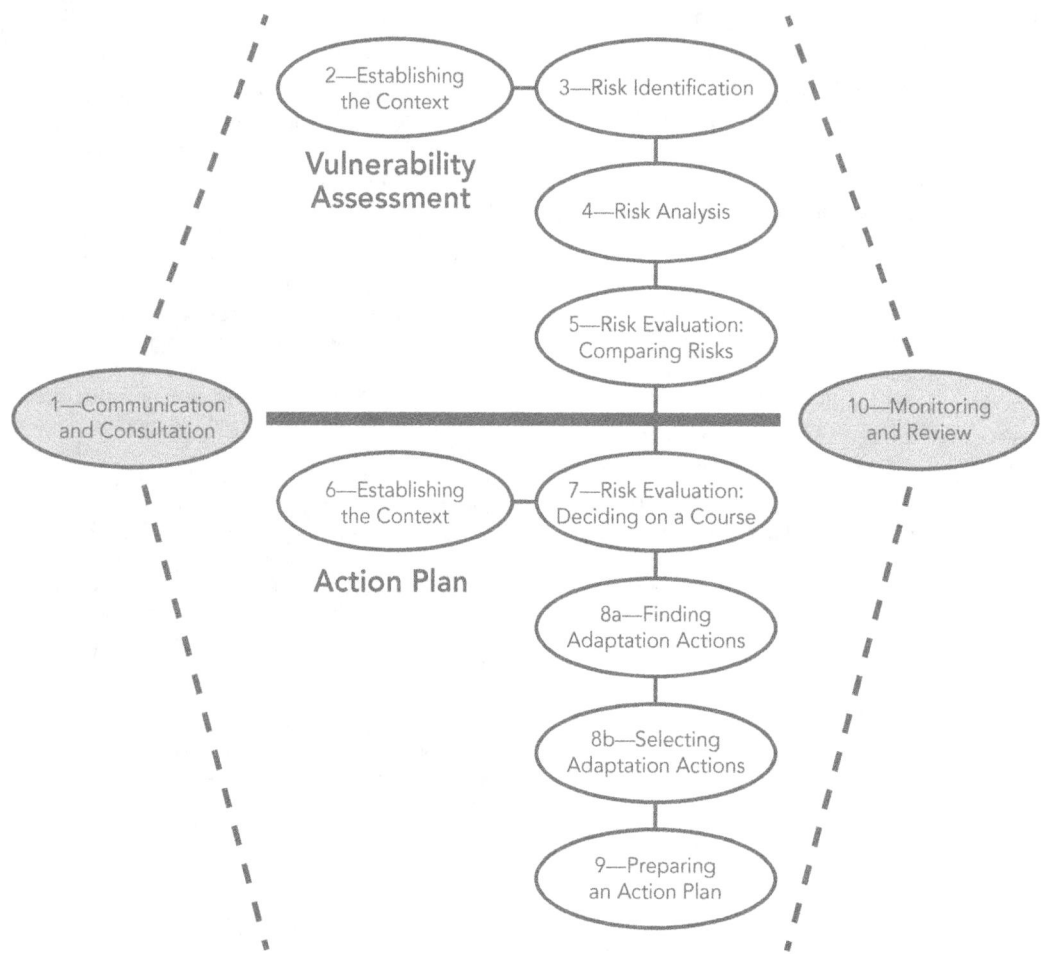

FIGURE I-1. A roadmap showing Step 1 through Step 5 of the vulnerability assessment and Step 6 through Step 10 of action planning. Communication and consultation should be part of every step. Monitoring and review keeps the whole adaptation plan up to date.

Vulnerability Assessments

A vulnerability assessment is an understanding of how climate change will uniquely affect your organization. The organization is the center of attention, and the reasons why it exists (i.e., its purpose, mission, goals or objectives) are the base units of the vulnerability assessment.

Organizational goals are the focus of concern. Goals are why organizations exist and what motivates them; goals provide a context for changes. The vulnerability assessment half of a risk-based adaptation plan is different from a climate change impacts study that describes how the future will be different if the climate is different. An impact that is identified without reference to a goal is just a detail. There is no context to say it is good or bad or that something should be done about it or not. Context comes from people or organizations. They can say that an impact will help, hurt or have no effect on what they are trying to do (Table I-1). If impacts will be unwanted, goals might still be attainable if risk-reducing actions can be implemented.

TABLE I-1. DIFFERENCES AMONG TYPES OF CLIMATE CHANGE PRODUCTS

	Climate change projection	Climate change impacts study	Risk-based vulnerability assessment
What is it?	A description of what future climate (and correlated phenomena such as sea level) will be like.	A description of how the future will be different if the climate is different.	A ranked description of how climate changes would keep an organization from reaching its goals.
Who creates it?	Climate modelers, assisted by scientists and social scientists who can provide the understanding or input necessary for scenarios or simulations.	Practitioners and experts who are qualified to describe how stressors would affect various human or natural systems.	Organizations, assisted by stakeholders, practitioners and other experts.
Example finding	The average annual temperature is expected to be 5°F warmer in 2100 than it was in 1990.	Warmer water provides better conditions for algal blooms and bacteria, and they are expected to become more plentiful.	There is a moderate risk that harmful algal blooms and higher bacteria levels from warmer water temperature will keep us from providing a safe place for people to swim.
Climate parameters	Climate projections are the calculated values for the modeled parameters.	Changing parameters become stressors.	Changing parameters become stressors.
Impacts	Not considered (unless a general awareness of impacts and a resulting change of globally significant behavior is incorporated into modeling scenarios).	Impact studies are a presentation about how things would be different because of climate change. Impacts could be described for a river, a forest, an island or a city. The study would describe what its future conditions would be like.	Knowledge of climate projections and future climate change impacts is a needed input.
Organization goals	Goals are not considered. Climate is the subject of modeling that tends to be done at a global scale or for areas that are much larger than any manager's responsibility.	Goals are not considered, although what someone considers to be important (whether stated or implied) will affect which impacts receive attention.	The introduction of an organization's goals is what differentiates a vulnerability assessment from an impacts study. The work can become the basis for action instead of a general report.
Risks	Not considered. This is largely about understanding the stressors.	Not considered because goals are necessary to have risks.	A vulnerability assessment is an evaluated set of risks that describes how climate change stressors would affect goals.
Risk-reducing actions	Not considered.	Does not have a basis for selecting or implementing actions.	The vulnerability assessment is input for choosing actions.

In using this WORKBOOK's risk management process for preparing a vulnerability assessment, you will take a broad look at all potential climate change risks, qualitatively assess them for their likelihood of occurrence and their consequences if they were to occur, and consult with your key stakeholders to gain buy-in and agreement on the assessment.

The WORKBOOK has five steps that lead to a climate change vulnerability assessment:

- **Step 1—Communication and Consultation**
  Informing key people about the vulnerability assessment and asking for input.

- **Step 2—Establishing the Context for the Vulnerability Assessment**
  Identifying organizational goals that are susceptible to climate change.

- **Step 3—Risk Identification**
  Brainstorming about how climate stressors will interact with your goals.

- **Step 4—Risk Analysis**
  Developing an initial characterization of consequence and likelihood for each risk.

- **Step 5—Risk Evaluation: Comparing Risks**
  Using a consequence/probability matrix to build consensus about each risk.

A vulnerability assessment represents a step forward, not an end in itself. However, the WORKBOOK treats a vulnerability assessment as a potential stand-alone product. A vulnerability assessment is a precursor to an action plan, but it will have value on its own for communication, coordination and decision support. A vulnerability assessment is a product to be used with your board of directors, senior management, and peer organizations, and with other levels of government. It is helpful for working with others to obtain funding, resources, buy-in or approval. Additionally, a broad look at all potential problems allows others to scan across places, compare neighboring systems, and find partners for addressing common risks. A vulnerability assessment is also a tool that will help you answer a series of questions about risks and mitigating options that you will need to ask when preparing your action plan.

While it is not necessary to have a stand-alone vulnerability assessment, there can be benefits to pausing after the assessment becomes available. Time to let the vulnerability assessment findings sink in may be useful. Pausing also provides an opportunity for consensus building before an action plan is developed. It is also a time to speak to potential partners about collaborating or dividing up tasks. A response that will effectively mitigate some risks to tolerable levels could be very expensive or politically sensitive. Securing resources that can be applied to climate change problems can also take time. Further, some risks might not start to cause problems for several more decades. For these reasons completing a vulnerability assessment could mark a temporary break in the adaptation process.

This is a risk management process for climate change. Taking a few days, weeks or months before launching into the action planning process could be useful. However, if there is no strategy to use an interim period or if the benefits of pausing are small, keep going with the action planning steps.

Action Plans

An action plan is a document that explains how you will manage your climate change risks. After completing a vulnerability assessment, you will have identified the risks that have the highest potential to affect your organization's ability to meet its goals.

The vulnerability assessment process helped you decide which risks to focus on. Now you will identify actions that might help to decrease those risks and increase the sustainability of your system. Making these decisions when you have limited resources and probably cannot mitigate all of your risks will not be easy. Pursuing win-win solutions or no-regrets actions will be appealing options that make sense regardless of what the future brings. **Steps 6–10** of the WORKBOOK lead to a climate change action plan:

- **Step 6—Establishing the Context for the Action Plan**
  Identifying opportunities and constraints that will affect your adaptation decisions.

- Step 7—Risk Evaluation: Deciding on a Course
  Deciding at a high level whether you will mitigate, transfer, accept or avoid each risk.

- Step 8a—Finding Adaptation Actions
  Finding mitigating actions that look promising for further investigation.

- Step 8b—Selecting Adaptation Actions
  Screening potential actions, and selecting a set of risk-reducing actions to implement.

- Step 9—Preparing and Implementing an Action Plan
  Creating a plan to track mitigating actions and which risks they address.

- Step 10—Monitoring and Review
  Keeping track of your actions and maintaining your vulnerability assessment.

This is a planning-level process to identify options. Much more detailed work will need to be done before any of the actions is ready to be implemented. The final steps in this WORKBOOK set up a process to track the status of your risks and to identify who will have the lead for various risk mitigation strategies.

Working with Others

In environmental management, it is impossible to be fully effective without engaging stakeholders. Involving others will pay off in many ways as you develop your adaptation plan. You will learn what topics potential partners are concerned about, who can contribute expertise, and whether you have support for moving forward. The process described in this WORKBOOK is well suited for a collaborative approach where key stakeholders can participate and work toward consensus. There is no one best way for communicating with stakeholders, and different methods will be appropriate in different steps of the WORKBOOK. WORKBOOK steps have lists of resources that can help with external communications. Appendix B also has cross-references to other guides that may be helpful.

Partnering with another organization in your area that has similar interests in assessing climate change can be a great opportunity. When you go through your communication and consultation process (Step 1), you should ask if your stakeholders want to join together to assess climate change effects.

If you decide to partner with another organization on a vulnerability assessment, Step 2—Establishing the Context for the Vulnerability Assessment provides an opportunity to identify common goals you will jointly consider. If the other organization is already a key partner, it is likely that there is significant overlapping interest and a common context can be established for this assessment.

During Step 3—Risk Identification and Step 4—Risk Analysis, working together to identify and assess risks can save resources or time, and lead to a common outcome. There can be great benefits to collaborating, and you certainly should when it makes sense. Partners might assess the same risk differently simply because their organizations have different purposes. Always remember to keep your organization's needs in mind.

In Step 6—Establishing the Context for the Action Plan and Step 7—Risk Evaluation: Deciding on a Course, partnering opportunities are an important part of the process. Identifying who has mutual interests and who can take the lead in mitigating particular risks will be a great way to accomplish more than any organization can do by itself.

In Step 9—Preparing and Implementing an Action Plan and Step 10—Monitoring and Review, let your partners know how you are doing. You will want to hear about their progress too.

Using the Workbook

Most of the information you need already exists. Some resources that will be helpful include the following:

- Your organization's management plan or strategic plan (for example, a National Estuary Program's Comprehensive Conservation and Management Plan).

- Knowledge of your organization's goals and objectives, key programs, and activities.

- Knowledge of your organization's history, operational procedures, and key stakeholders or partners.

- Maps of your watershed or study area.

- Any existing assessments or climate readiness work that covers your study area. Regional and sectoral studies from the National Climate Assessment can be useful.

EPA's *Handbook for Developing Watershed Plans to Restore and Protect Our Waters* (Chapter 5, pp. 5-1 through 5-49, http://water.epa.gov/polwaste/nps/handbook_index.cfm) has recommendations for where to find and how to gather existing data.

It is important to follow the risk management steps in order. Each step builds on the previous one in this systematic process (Appendix A). Skipping ahead will lead to an incomplete or imperfect outcome, but you can always go back and repeat previous steps as more information becomes available.

Each step in the WORKBOOK contains the following sections:

- What Is [Step x].
- Objective of This Step.
- Process.
- To Get Started.
- A document to develop or a table to fill out.
- Additional Resources.

The tables and documents that you produce as you go through the WORKBOOK are a record of your process and decisions. The tables from Step 2– Step 9 are progressive: information generated in one step informs the steps that follow. Keeping track of your contacts and sources will make it much easier if later you need to inform your stakeholders or reconsider your conclusions. The record you develop during the vulnerability assessment will be very useful when it is time to develop an action plan because a major part of the decision-making process will be a consideration of how to reduce the vulnerability of your system. Knowing how you reached your determinations about the likelihoods and consequences of scores of risks will be the key to reducing them.

Cross-reference to climate change planning you may have already completed

If your organization, community, region or state has already used NOAA's *Roadmap for Adapting to Coastal Risk*, ICLEI's *Preparing for Climate Change—A Guidebook for Local, Regional, and State Governments*, or NOAA's *Adapting to Climate Change: A Planning Guide for State Coastal Managers*, the information generated in that work will be helpful to you in conducting this vulnerability assessment. In Appendix B, sections of the NOAA and ICLEI processes are cross-referenced to identify helpful information and to assist those who have already worked with those resources.

Michael Craghan, EPA Office of Water

What Is "Communication and Consultation"?

As you begin, communication and consultation is an opportunity to inform your key stakeholders and partners about why a climate change adaptation plan is necessary, as well as to describe the process and the expected products. Consulting with stakeholders helps to build support for adaptation and is essential in developing agreement on the outcome of this process. It is also an opportunity to gain knowledge and information from them.

Objective of This Step

The objective of this step is to list your key stakeholders and learn their particular interests or concerns about climate change risks and the adaptation planning process. You will also prepare a schedule for stakeholder involvement.

Process

You should reach out to others who are interested in what you are doing or who can help with your adaptation planning. Communicate the purpose of this work to decision-makers within your organization and to key stakeholders. Everyone should understand what you are trying to accomplish and the expected outcomes.

Be able to describe what happens in each of the steps before reaching out. What you tell and ask other people will be much more useful if you know how the rest of the process will unfold.

Some may be involved throughout the entire process, while others may have a particular interest in a single step or area of focus. However, when Step 5 is completed, key people should not be surprised to learn that you have a climate change vulnerability assessment and that you will be using that assessment to determine what further actions are necessary.

Reaching out

Every organization has a unique group of stakeholders. Local governments, nonprofits, federal agencies or watershed groups all have different structures and different parties that pay attention to what they do. Who is involved in climate change adaptation planning is an organizational decision. Leading candidates would include:

> **Stakeholder involvement in the National Estuary Program**
>
> In Section 320 of the Clean Water Act, Congress said that the members of [an NEP's] initial management committee should include (in addition to states and government agencies), "local governments," and "affected industries, public and private educational institutions, and the general public…" Stakeholder engagement is baked right into an NEP.
>
> Involving stakeholders will lead to a better assessment and wider public support. In *Community-Based Watershed Management: Lessons from the National Estuary Program* (pp. 13–14), EPA says that members of a citizen's advisory committee generally meet one or more of these criteria:
>
> - Serve as spokespersons for a major user or interest group and bring information back to that group.
> - Are well-respected leaders in the community.
> - Have experience in the development of water quality and resource management policy.
> - Have experience with volunteer nonprofit groups, the general public, outreach and education activities, and the media.
> - Understand the technical and economic feasibility of the pollution control options under consideration.
> - Understand the consensus-building process.
> - Are energetic and motivated individuals.

- anyone who would be part of organizational strategic planning;
- anyone who would help decide an annual work plan;
- anyone who could help with the vulnerability assessment; and
- anyone who might be able to help with implementing adaptation actions.

Including the public in your planning can be helpful. The "Additional Resources" section in this step has many resources about how the community can help, as well as strategies for engaging with them.

Communication and consultation should occur throughout your planning, not just when it starts and ends. Actively review the information developed here in this step and listed in Table 1-1 before each subsequent step and update it throughout the process to track and guide communication and consultation efforts.

Key messages to communicate at the outset

It will not be possible to explain to others what the conclusions of the vulnerability assessment are going to be. However, it will be possible to explain what you will be doing, how conclusions will be reached, and what could be the next steps. Along with the organizational goals and objectives you are considering for this assessment, key messages should include the following:

- This vulnerability assessment is being launched because there is reason to believe that climate change impacts will affect what your organization is trying to accomplish.
- The purpose of the vulnerability assessment is to understand the risks that may occur, the likelihood of occurrence for each risk, and the consequence if a risk should occur.
- Risk management is part of decision-making. If climate changes are going to affect your organization's ability to reach its goals, then you need to know what the impacts might be in order to continue being effective.
- Different perspectives on risk identification, risk analysis and risk evaluation are important, and participation is welcome.
- The outcome of the vulnerability assessment is an understanding of likely climate change risks; any adaptation actions will be determined after that.

Key questions for your consultations with stakeholders

By nature, your stakeholders are invested and concerned with what you are doing. They can be a great help as you develop and ultimately implement your adaptation plan. Let them know how you plan to use their input. Your communication and consultation should strive to find answers to key questions about how stakeholders can improve your vulnerability assessment or action plan:

- Do they have resources, reports or data that they can contribute?
- Do they know of any climate change work on stressors or impacts that exists or is ongoing?
- Do they want to actively participate in the process or simply be informed of progress or updates?
- Can they help with risk identification or risk analysis?
- When it is time to decide on adaptation actions, would they be able to help implement actions to reduce risks?
- Do they have any "no-go"s that you should know about? Are there any topics that they are not willing to discuss or consider?

To Get Started

The first group of people to turn to is your organization's key personnel: your board of directors, management, staff, and the people who are regularly involved with what you do.

TABLE 1-1. STAKEHOLDER INVOLVEMENT

Stakeholder	Interests/concerns	Level of involvement	During which step(s) should you reach out to them?
1.		☐ Not participating ☐ Stay informed ☐ Active participant	
2.		☐ Not participating ☐ Stay informed ☐ Active participant	
3.		☐ Not participating ☐ Stay informed ☐ Active participant	
n.		☐ Not participating ☐ Stay informed ☐ Active participant	

Additional Resources

Also see Appendix B.

EPA resources for involving the community

EPA. 2010. *Getting In Step: A Guide for Conducting Watershed Outreach Campaigns.* 3rd ed.
http://www.epa.gov/owow/watershed/outreach/documents/getnstep.pdf

EPA. 2013. *Getting in Step: Engaging and Involving Stakeholders in Your Watershed.* 2nd ed.
http://cfpub.epa.gov/npstbx/files/stakeholderguide.pdf

EPA. 2008. *Handbook for Developing Watershed Plans to Restore and Protect Our Waters.*
http://water.epa.gov/polwaste/nps/handbook_index.cfm

EPA. 2005. *Community-Based Watershed Management: Lessons from the National Estuary Program (NEP).* Chapter 3, pp. 11–15.
http://water.epa.gov/type/oceb/nep/upload/2007_04_09_estuaries_nepprimeruments_NEPPrimer.pdf

EPA. 2002. *Community Culture and the Environment: A Guide to Understanding a Sense of Place.*
http://www.epa.gov/care/library/community_culture.pdf

Climate Ready Water Utilities. *Climate Ready Water Utilities Workshop Planner.*
http://water.epa.gov/infrastructure/watersecurity/climate/

Resources for engaging stakeholders

NOAA Coastal Services Center. 2007. *Introduction to Stakeholder Participation.*
http://csc.noaa.gov/digitalcoast/publications/stakeholder

National Wildlife Federation. 2011. *Scanning the Conservation Horizon—A Guide to Climate Change Vulnerability Assessment.* Chapter 2, p. 32.
http://www.nwf.org/~/media/PDFs/Global-Warming/Climate-Smart-Conservation/NWFScanningtheConservationHorizonFINAL92311.ashx

USDA U.S. Forest Service. 2011. *Responding to Climate Change in National Forests: A Guidebook for Developing Adaptation Options.* pp. 22-28.
http://www.treesearch.fs.fed.us/pubs/39884

Cooperative Institute for Coastal and Estuarine Environmental Technology and Wells Reserve NERR. 2008. *Collaborative Learning Guide for Ecosystem Management.*
http://www.wellsreserve.org/sup/downloads/collaborative_learning_guide.pdf

Eric Vance, EPA

What Is "Establishing the Context for the Vulnerability Assessment"?

In this step you will explicitly identify your organization's goals. These goals will establish the scope and boundaries for the vulnerability assessment process. This information will help keep you focused: climate changes that do not affect your organization's goals would not be part of your assessment.

Objective of This Step

The objective of this step is to find and list your organizational goals.

Process

Every organization exists for a reason. Your charter or strategic plan lists these purposes. In this step you are not being asked to generate these purposes; all you need to do is find your already existing goals and formally incorporate them into the vulnerability assessment process. List these goals in Table 2-1.

Your organization is the center of attention

Your vulnerability assessment is specifically about your organization. It is not about your region or your place. Your organization, like all the others that work there, was created to fulfill some responsibility or to accomplish something that needed attention. You have a mission that describes what you strive for and why you are unique. Your organization has a role in your community and that role is why your patrons, partners and supporters look to you.

A characteristic of environmental management is to look holistically at a place. However, right now you need to focus on your niche within it. Many climate change impacts in your location will not affect why you exist or what you work on. You don't need to take on every challenge that climate change may bring to your place.

The purpose of a climate change adaptation plan is to make sure you can continue to achieve the organizational goals to accomplish your mission. Essentially you are reviewing your organization's already-existing goals to see how climate change may threaten their achievement. A vulnerability assessment that is attentive to your organization will help you write an action plan that will keep you on track. You are not setting out to write a plan for your whole region (unless that is part of your responsibility).

How others fit in

At the watershed scale for which the WORKBOOK is designed, it would be common to find government agencies, communities, nonprofits and others who care about the same things and share your organizational goals. If two organizations are common partners, their goals are probably similar enough that they could work together on parts of a vulnerability assessment. They might need to evaluate the consequences of the risks to their respective organizations differently because they have different contexts and different missions, but risk identification and other elements of the risk analysis could be done jointly. These organizations will probably want to coordinate later about action plans related to those goals as well.

Goals are the fundamental elements

Goals are the natural unit for seeing what problems climate change may present and therefore they have important roles in the vulnerability assessment.

Goals describe what your organization intends to achieve, and are therefore perfect for defining the scope of your vulnerability assessment. Your vulnerability assessment should include everything that affects your goals, while omitting everything that has no effect on them. A useful

> ### Goals
>
> "Strategic goals should reflect the broad, long-term, outcomes the agency aspires to achieve by implementing its mission. Strategic goals communicate the agency efforts to address national problems, needs, challenges, and opportunities on behalf of the American people. Both the way strategic goals are framed and the substance they communicate are important to consider. Strategic goals should reflect the statutory mission of the agency, and most agency activity will align to the strategic goals. Strategic goals need not be as specific as strategic objectives, however, and need not reflect every activity that the agency must undertake to accomplish its mission."
>
> OMB. 2013. Circular No. A–11 (2013), Section 230—Agency Strategic Planning.

test that keeps your task from getting ever larger is to ask whether you would allocate resources for efforts that have no effect on your goals. Your assessment should focus on the things your organization cares about. You have already articulated your priorities when setting your goals.

In addition, if there is no goal to test against climate stressors, then there will be no risks. The plain detail that a stressor would cause something to occur in a place does not have meaning within itself. Once it could affect a goal, an impact takes on a consequence. For example: you might identify that it could rain tomorrow—but that would be just an incident; for many people that information would have no value. However if you say it could rain tomorrow and we might not be able to have the picnic we planned, then you introduce a consequence and you have a risk. A goal is the necessary piece for naming risks. There will be climate change impacts all over your region but risks can only emerge if you cross climate change stressors with your goals.

Your organization may have goals that would be affected by climate change (e.g., controlling water pollution or maintaining habitat) but may have others (e.g., recycling waste) that would not be affected much at all. You will have to decide which of your organization's goals will be considered in this vulnerability assessment. For a planning-level study like this, you should move forward with all of those that could be affected by climate change.

Clean Water Act goals

If your organization's goals fit under the Clean Water Act umbrella, then the risk identification checklist for those themes in **Step 3** may be useful when you reach that step.

Purposes in Clean Water Act §320, and the Estuaries and Clean Waters Act of 2000:

- Control point and nonpoint sources of pollution and clean up pollution.
- Maintain and improve estuarine habitat.
- Protect and propagate fish, shellfish and wildlife, including control of nonnative species.
- Protect public water supplies and recreational activities, in and on the water.

To Get Started

Turn to your organization's strategic plan, management plan or charter, which should identify your mission, goals, objectives, or the issues your organization cares most about.

TABLE 2-1. ORGANIZATIONAL GOALS

Goal	Does it correspond with one of the Clean Water Act purposes? (Y/N)
1.	
2.	
3.	
n.	

Additional Resources

Also see Appendix B.

EPA resources about goal setting and strategic planning

EPA. 2005. *Community-Based Watershed Management: Lessons from the National Estuary Program.*
http://water.epa.gov/type/oceb/nep/upload/2007_04_09_estuaries_nepprimeruments_NEPPrimer.pdf

EPA. 1992. *National Estuary Program Guidance: Comprehensive Conservation and Management Plans: Content and Approval Requirements.*
http://nepis.epa.gov/Exe/ZyPURL.cgi?Dockey=20004XHU.txt

EPA. *Climate Ready Water Utilities (CRWU).*
http://water.epa.gov/infrastructure/watersecurity/climate/

John Fleck, FEMA

What Is "Risk Identification"?

This is the process of generating a broad list of reasonably foreseeable ways that climate change stressors could keep your organization from achieving its goals. It is important to consider all potential risks during the risk identification step. If risks are not identified in this step, they will not be analyzed and evaluated in the steps that follow.

Objective of This Step

The objective of this step is to create a broad list of climate change risks that might affect your organization's ability to achieve its goals.

Process

The work in this step involves using your understanding of your place to identify risks by thinking about how climate change stressors would interact with your organization's goals. Compile a list of all of your risks in Table 3-3 using information in the checklists of Tables 3-1a through 3-1d and anything you add to Table 3-2.

What is a "risk"?

Risks threaten things that are of value. In the context of climate change, a risk is the possibility that a given climate change stressor will affect your organization's ability to meet its goals. A risk is a problem to be managed by finding ways to lower its principal characteristics: likelihood and consequence. In this step, you will cross your goals with climate change stressors to identify risks.

Climate change stressors

The WORKBOOK uses seven types of climate change stressors to organize thinking:

- Warmer summers (overall climate)
  This stressor is generally about the warm season being even warmer. This stressor (like warmer winters, below) is about the general climate. Air, surface, soil and groundwater temperatures will be warmer. The general climate effects of having warmer oceans or lakes are included here.

- Warmer winters (overall climate)
  This stressor is about a cold season not being as cold as it formerly was.

- Warmer water
  This stressor (regardless of season) comes from a higher temperature of water bodies (including the ocean) and affects the chemical, physical or biological characteristics of the water body itself.

- Increasing drought
  Drought is a deficiency in precipitation over an extended period. The magnitude of the deficiency, the duration or the number of droughts could be greater.

- Increasing storminess
  This category encompasses all aspects of intensifying precipitation in any form: more seasonal precipitation, more total precipitation during events, higher rates of precipitation during events. Stronger or more frequent instances of extratropical and tropical cyclones, blizzards or other weather conditions are included here. If they are acting as stressors, then floods, waves, coastal storm surge and wind are part of this storminess category.

- Sea level rise
  This stressor is about the ocean being higher than it formerly was. It includes effects of higher water levels right at the shore, as well as how elevated coastal water levels affect inland systems.

- Ocean acidification
  For the WORKBOOK, this category is primarily conceptualized as related to ocean acidification via atmospheric inputs of carbon dioxide.

If you find that your place is impacted by an additional climate stressor that is not listed above, then add it to your set. Conversely, for your risk identification, if a stressor is not relevant to what your organization does, or if there are good reasons to believe the impact it is associated with would not matter, then omit it.

Risk paths

Recall that a risk is the possibility that a given climate change stressor will affect your organization's ability to meet its goals. Stressors and goals are embedded in the risk. You could be prevented from reaching a goal because the environment changes in some way or because the projects you undertake to reach your goal would fail in some way.

If there is any potential sequence (climate change stressor, and what follows) for an unwanted consequence (not reaching your goal), then you have a risk. The risk develops along the path between the cause and the effect. For example:

- Stressor X could _____, and the result is that we might not attain Goal Y.
- Warmer water (stressor) could lead to bacteria being more abundant in the river (path), and we might not be able to provide healthy drinking water (unrealized goal).
- More intense precipitation (stressor) could lead to flooding that knocks a sewage pump station offline (path) and we might not be able to treat all the sewage that is discharged to the bay (unrealized goal).
- Sea level rise (stressor) will lead to more beach erosion (path) and we might not be able to maintain the endangered bird species nesting sites (goal).

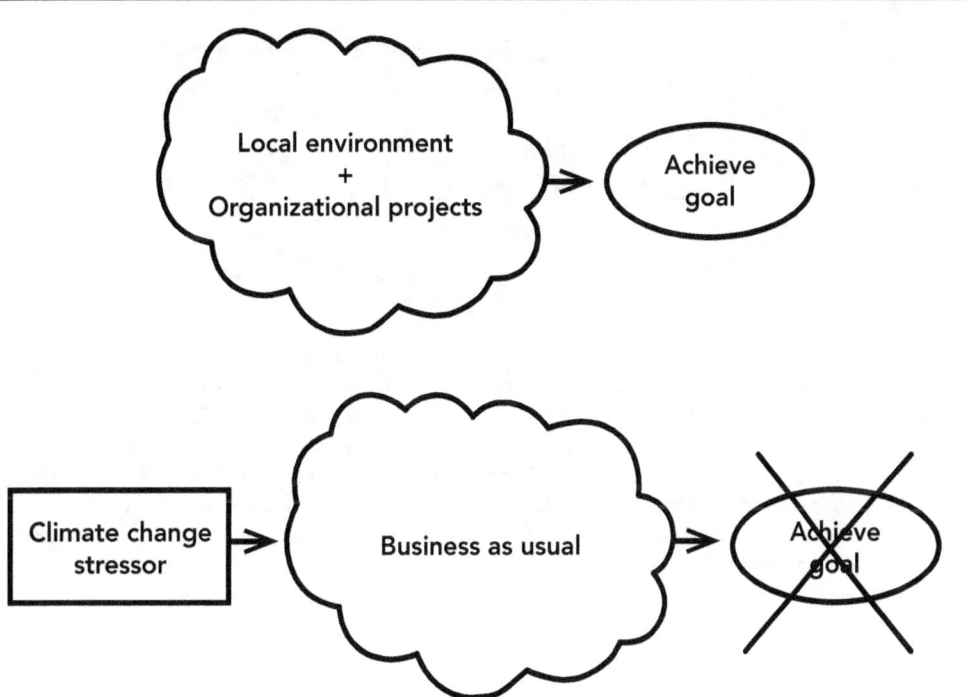

FIGURE 3-1. The top diagram depicts an ideal business as usual scenario. At your place, your organization undertakes various projects that help you to achieve your goals. In the bottom diagram a climate change stressor is introduced to the business as usual situation (or to a no-action scenario). The stressor could change the way the environment functions or change the effectiveness of the projects that you would undertake, and potentially lead to unrealized goals. After the stressor is introduced, if you can describe some chain of events that could happen that might keep you from realizing a goal, then you have a risk.

Paths can be simple or complex. Paths might involve contingencies or future human activity.

A diagram of a system can be very helpful for visualizing how climate change stressors could alter its operations. It is not necessary to have climate change stressors in the diagram; the illustrations are just a tool to help you think about where stressors could introduce unwanted effects. Diagrams can be simple sketches of sequential events (Figure 3-2). If you can easily locate a conceptual diagram (Appendix D) for a system comparable to yours, then a tool like that can be useful as well. Diagrams do not have to be comprehensive or even exact representations of reality. They are simply tools to help you brainstorm and identify risks.

Daily Operations
Contribute to Success

Most strategic initiatives are built on existing processes, and much of the organization is involved in the day-to-day work of delivering service to our customers. The Postal value chain is a series of closely linked processes, supported by a common infrastructure.

The overall quality and cost of the system depends not only on the actions of the Postal Service, but also on the actions of mailers and mail service providers.

Value Chain

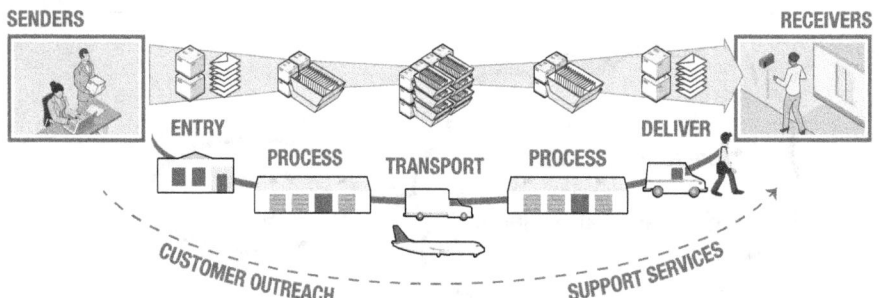

U.S. Postal Service

FIGURE 3-2. A visual depiction of the current system helps with identifying risks. Climate change stressors like increasing storminess or hotter summers can be imagined to affect efficient mail delivery in many ways. A diagram like this, from the USPS *FY 2012 Annual Report to Congress*, helps prompt questions such as "How would more wind and rain affect maintenance at post offices?" "Would hotter conditions require more air conditioning?" "Does increasing temperature affect vehicle or aircraft operations"? "Would more intense precipitation or more heat alert days affect mail delivery?" You might have come up with these risks even without an illustration—but if you can have a diagram that helps you think, then use it.

What you need to do

Every path that you can conceive is a risk to a goal and should be recorded in this risk identification step. At this point, you need to rely on your expertise and knowledge about your system. In this step you want to come up with as many of those possible pathways as you can.

Cross each climate change stressor with each of the goals you listed in Step 2. The intent is to generate a list of the ways that each stressor could keep you from reaching the goal. This is much like a brainstorming session. See Figure 3-3 for an example of how the process works for one goal and one stressor.

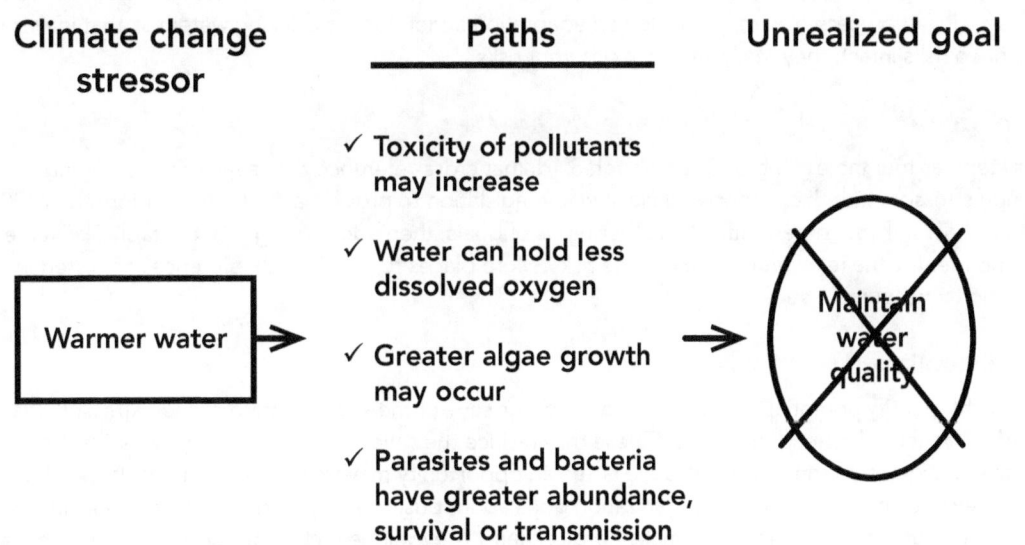

Climate change stressor — **Paths** — **Unrealized goal**

Warmer water →

✓ Toxicity of pollutants may increase

✓ Water can hold less dissolved oxygen

✓ Greater algae growth may occur

✓ Parasites and bacteria have greater abundance, survival or transmission

→ Maintain water quality (crossed out)

FIGURE 3-3. An example of how to generate a list of what could happen in the space between introduction of a stressor and failure to realize a goal.

Record each risk (stressor, path and goal) in Table 3-3. If a potential risk is not identified in this step, it will not move forward in the vulnerability assessment process, so write down everything. Step 4—Risk Analysis and Step 5—Risk Evaluation: Comparing Risks are where you analyze and evaluate the list of risks you generate in this step.

Don't dismiss risks!

Even risks that you might think are insignificant should be captured in this step. Don't prematurely dismiss anything. You may learn that a stakeholder thinks it is important, or in some instances further analysis might show that a risk is not trivial at all. If you dismiss it now, it does not get considered later. If it is in fact a trivial risk, the systematic process of risk management allows it to be assigned to such a category in Step 4. Treat this step as a brainstorming session and document every risk you and your team can identify.

Risks are not inevitable

A risk is not certain to happen, a risk could happen. You will analyze its likelihood in Step 4. Further, if you decide that you want to change its likelihood, understanding the risk path could help you to find ways to take action. You probably cannot—by yourself—change a stressor coming from global climate change, and you probably don't want to change your goal, so if you do want to change the likelihood or consequence of the risk you have to disrupt the path that transfers the risk from stressor to goal.

More than one problem from the same cause

The same environmental stressor may intersect with several of your goals. You may find that you seemingly identify the same risk many times. As an example, sea level rise may push saltwater farther up rivers. This could have implications for drinking water sources, freshwater wetlands or shellfish habitat. This is why you are asked to note which organizational goal is associated with a risk. Goals are implicitly embedded in risks. In this sea level rise example, you really haven't identified the same risk of saltwater reaching farther upstream three times: you have identified three separate risks that stem from the same stressor-driven process. You might design an adaptation action for the drinking water risk that leaves the other two risks intact: they really are three different risks.

Help identifying risks

This step has four tables (Tables 3-1a through 3-1d) that cross-reference the seven types of climate change stressors with the purposes of clean water legislation to provide a starter list of more than 100 potential risks. If your organization has other types of goals, then add those goals to a table like Table 3-2 and identify the relevant climate change risks. More places to turn for risk references are listed in the "Additional Resources" section.

Opportunities (possibility of benefits)

You might identify potential circumstances arising from any of the seven climate change stressors that could be beneficial instead of harmful. These may reduce the severity of some other risk or just be positive outcomes on their own. If it is the latter, an opportunity may free resources to address other risks or even change the way your organization goes about business. If you identify any opportunities related to your organization's goals, make note of them. They will be picked back up in your action plan to ensure that you take advantage of them.

TABLE 3-1A. POTENTIAL CLIMATE CHANGE RISKS FOR POLLUTION CONTROL

Clean Water Act goals	Warmer summers	Warmer winters	Warmer water	Increasing drought	Increasing storminess	Sea level rise	Ocean acidification
Controlling point sources of pollution and cleaning up pollution		☐ Loss of melting winter snows may reduce spring or summer flow volume, and raise pollutant concentration in receiving waters	☐ Temperature criteria for discharges may be exceeded (thermal pollution) ☐ Warmer temperatures may increase toxicity of pollutants	☐ Critical-low-flow criteria for discharging may not be met ☐ Pollutant concentrations may increase if sources stay the same and flow diminishes	☐ Combined sewer overflows may increase ☐ Treatment plants may go offline during intense floods	☐ Treatment plants may not be able to discharge via gravity at higher water levels ☐ Treatment infrastructure may be susceptible to flooding ☐ Sewage may mix with seawater in combined sewer systems ☐ Contaminated sites may flood or have shoreline erosion ☐ Sewer pipes may have more inflow (floods) or infiltration (higher water table)	
Controlling nonpoint sources of pollution	☐ Wildfires may lead to soil erosion	☐ Longer growing season can lead to more lawn maintenance with fertilizers and pesticides	☐ Higher solubility may lead to higher concentration of pollutants ☐ Water may hold less dissolved oxygen ☐ Higher surface temperatures may lead to stratification ☐ Greater algae growth may occur ☐ Parasites, bacteria may have greater survival or transmission	☐ Pollution sources may build up on land, followed by high-intensity flushes	☐ Streams may see greater erosion and scour ☐ Urban areas may be subject to more floods ☐ Flood control facilities (e.g., detention basins, manure management) may be inadequate ☐ High rainfall may cause septic systems to fail	☐ Tidal flooding may extend to new areas, leading to additional sources of pollution	☐ Decomposing organic matter releases carbon dioxide, which may exacerbate the ocean acidification problem in coastal waters

TABLE 3-1B. POTENTIAL CLIMATE CHANGE RISKS FOR HABITAT

Clean Water Act goals	Warmer summers	Warmer winters	Warmer water	Increasing drought	Increasing storminess	Sea level rise	Ocean acidification
Restoring and protecting physical and hydrologic features	□ Higher temperatures may lead to greater evaporation and lower groundwater tables □ Switching between surface and groundwater sources for public water supplies may affect the integrity of water bodies □ Greater electricity demand may affect operation decisions at hydropower dams	□ Less snow, more rain may change the runoff/infiltration balance; base flow in streams may change □ A spring runoff pulse may disappear along with the snow □ Rivers may no longer freeze; a spring thaw would be obsolete □ Marshes and beaches may erode from loss of protecting ice	□ Warmer water may lead to greater likelihood of stratification	□ Groundwater tables may drop □ Base flow in streams may decrease □ Stream water may become warmer □ Increased human use of groundwater during drought may reduce stream baseflow □ New water supply reservoirs may affect the integrity of freshwater streams	□ The number of storms reaching an intensity that causes problems may increase □ Stronger storms may cause more intense flooding and runoff □ Coastal overwash or island breaching may occur □ Turbidity of surface waters may increase □ Increased intensity of precipitation may yield less infiltration	□ Shoreline erosion may lead to loss of beaches, wetlands and salt marshes □ Saline water may move farther upstream and freshwater habitat may become brackish □ Tidal influence may move farther upstream □ Bulkheads, sea walls and revetments may become more widespread	
Constructing reefs to promote fish and shellfish			□ Desired fish may no longer be present □ Warmer water may promote invasive species or disease		□ Stream erosion may lead to high turbidity and greater sedimentation □ Lower pH from NPS pollution may affect target species	□ Light may not penetrate through deeper water □ Higher salinity may kill targeted species	□ Long-term shellfish sustainability may be an open question □ Fish may be adversely affected during development stages

TABLE 3-1C. POTENTIAL CLIMATE CHANGE RISKS FOR FISH, WILDLIFE AND PLANTS

Clean Water Act goals	Warmer summers	Warmer winters	Warmer water	Increasing drought	Increasing storminess	Sea level rise	Ocean acidification
Protecting and propagating fish, shellfish and wildlife	☐ Species that won't tolerate warmer summers may die/migrate; biota at the southern limit of their range may disappear from ecosystems	☐ Species that used to migrate away may stay all winter ☐ Species that once migrated through may stop and stay	☐ Newly invasive species may appear ☐ Habitat may become unsuitably warm, for a species or its food ☐ Heat may stress immobile biota ☐ Dissolved oxygen capacity of water may drop	☐ Species may not tolerate a new drought regime ☐ Native habitat may be affected if freshwater flow in streams is diminished or eliminated	☐ Greater soil erosion may increase turbidity and decrease water clarity ☐ Greater soil erosion may increase sediment deposition in estuaries, with consequences for benthic species	☐ Sea level may push saltier water farther upstream (especially of interest with regard to shellfish habitat) ☐ Light may not penetrate through the full depth of deeper water ☐ Greater coastal wetland losses may occur	☐ Corrosive waters may impact shellfish development ☐ Shellfish predators may not survive the disappearance of shellfish ☐ Fish may be adversely affected during development stages by changes to water chemistry ☐ The effect of ocean acidification on calcifying plankton may lead to cascading effects in the food chain
Controlling nonnative and invasive species	☐ Species may be weakened by heat and become out-competed	☐ Pests may survive winters that used to kill them ☐ Invasive species may move into places that used to be too cold	☐ Some fish reproduction may require cold temperatures; other reproductive cycles are tied to water temperature	☐ Changing freshwater inputs may affect salinity distribution in estuaries (especially of interest with regard to shellfish habitat)			
Maintaining biological integrity and reintroducing native species	☐ Essential food sources may die off or disappear, affecting the food web ☐ Species may need to consume more water as temperature rises	☐ Some plants may need a "setting" cold temperature ☐ A longer growing season may lead to an extra reproductive cycle ☐ Food supplies and bird migrations may be mis-timed	☐ Coral bleaching episodes may increase ☐ Parasites and diseases are enhanced by warmer water				

TABLE 3-1D. POTENTIAL CLIMATE CHANGE RISKS FOR RECREATION AND PUBLIC WATER SUPPLIES

Clean Water Act resource goals	Warmer summers	Warmer winters	Warmer water	Increasing drought	Increasing storminess	Sea level rise	Ocean acidification
Restoring and maintaining recreational activities, in and on the water	☐ More people using water for recreation may raise the potential for pathogen exposure		☐ Harmful algal blooms may be more likely ☐ Jellyfish may be more common ☐ Fishing seasons and fish may become misaligned ☐ Desired recreational fish may no longer be present ☐ Invasive plants may clog creeks and waterways	☐ Freshwater flows in streams may not support recreational uses ☐ Increased estuary salinity may drive away targeted recreational fish	☐ More frequent or more intense storms may decrease recreational opportunities ☐ Greater NPS pollution may impair recreation	☐ Beaches or public access sites may be lost to coastal erosion or inundation ☐ Clearance under bridges may decrease	☐ Eco-tourism resources or attractions (e.g., birding, diving, fishing) may be degraded ☐ Recreational shellfish harvesting may be lost
Protecting public water supplies	☐ Warmer temperatures may drive greater water demand ☐ Evaporation losses from reservoirs and groundwater may increase	☐ Summer water supplies that depend on winter snow pack may disappear ☐ Cold places may see more freeze/thaw cycles that can affect infrastructure	☐ Changes in treatment processes may be required ☐ Increased growth of algae and microbes may affect drinking water quality	☐ Lower freshwater flows may not keep saltwater downstream of intakes ☐ Groundwater tables may drop ☐ Coastal aquifers may be salinized from insufficient freshwater input ☐ Coastal aquifers may be salinized from higher demand on groundwater ☐ Maintaining passing flows at diversions may be difficult	☐ Water infrastructure may be vulnerable to flooding ☐ Flood waters may raise downstream turbidity and affect water quality	☐ Sea level may push salt fronts upstream past water diversions ☐ Water infrastructure may be vulnerable to inundation or erosion ☐ Saltwater intrusion into groundwater may be more likely	

TABLE 3-2. ADDITIONAL GOALS

Organizational goals	Warmer summers	Warmer winters	Warmer water	Increasing drought	Increasing storminess	Sea level rise	Ocean acidification
Goal ___							
Goal ___							
Goal ___							

To Get Started

Take a look at the climate change risks listed in Tables 3-1a through 3-1d. Check off each risk that applies to your organization's goals. It is conceivable that most of these risks will be checked.

TABLE **3-3. R**ISKS

Organizational goal	Climate stressor	Risk	Is this an opportunity instead of a risk?
1.			
2.			
3.			
n.			

Additional Resources

Also see Appendix B.

EPA resources about climate change risks for water resources

EPA Region 9 and California Department of Water Resources. 2011. *Climate Change Handbook for Regional Water Planning.*
http://www.water.ca.gov/climatechange/CCHandbook.cfm

Climate Ready Water Utilities. 2012. *Adaptation Strategies Guide for Water Utilities.*
http://water.epa.gov/infrastructure/watersecurity/climate/upload/epa817k11003.pdf

Coastal management goals

NOAA. 2010. *Adapting to Climate Change: A Planning Guide for State Coastal Managers.* Chapter 2.
http://coastalmanagement.noaa.gov/climate/docs/adaptationguide.pdf

Resources for identifying climate sensitivities for estuaries

NOAA. 2013. *Climate Sensitivity of the National Estuarine Research Reserve System.*
http://nerrs.noaa.gov/Doc/PDF/Research/130725_climate%20sensitivity%20of%20nerrs_Final-Rpt-in-Layout_FINAL.pdf

Nancy Laurson, EPA Office of Water

What Is "Risk Analysis"?

Risk analysis is the process of understanding a risk, which includes being aware of the driving force of the risk, assessing the likelihood (probability) of it occurring, and assessing the consequence if it were to occur. Risk analysis is essential to making decisions about which risks will become organizational priorities.

Objective of This Step

The objective of this step is to make an initial, high-level determination of the consequence, likelihood, spatial scale of the impact, and the time horizon until a problem begins for the climate change risks you identified in Step 3, so they can be sorted into high-medium-low qualitative categories of impact.

Process

This is the most intensive step of the vulnerability assessment. You will use Table 4-1 to characterize each risk in five areas: (1) consequence, (2) likelihood, (3) spatial extent of the impact, (4) time horizon until the problem begins, and (5) habitat type. This will become important information when your organization determines its adaptation priorities.

Each risk needs to be assigned to the minimum number of people or teams (ideally just one) who can provide a reliable initial analysis. You need an initial analysis of each risk along with some documentation of the sources (which could be expert judgement). Try not to get stuck on determining the likelihood or consequence of any one risk. Right now this will be an initial or a working determination. As more information becomes available over time, or as others are able to assist, you can return to this step and adjust the determination accordingly.

Whereas the WORKBOOK methodology brings others into the process in Step 5, it could be desirable to include others now if stakeholders indicated that they would help you with this part of your vulnerability assessment. The involvement of more than one person can lead to differences in whether risks are rated high, medium or low. The intent of this step is to provide one initial rating so that it can be refined via wider consultation and participation in Step 5—Risk Evaluation: Comparing Risks. If you do receive divergent input in this step, you need to come to a conclusion about which rating you will provisionally adopt. Differences could be worked out here or postponed to Step 5.

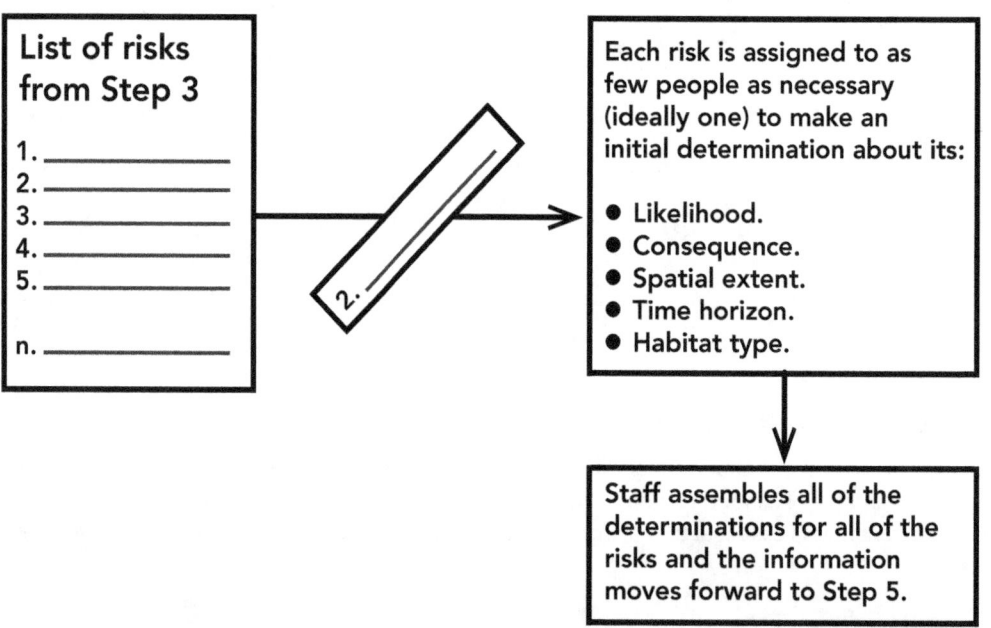

FIGURE 4-1. In this step, each risk from the list compiled in Step 3 is separately assigned to someone (or to a team) who is qualified to come up with qualitative determinations of likelihood, consequence, spatial extent, time horizon or habitat type. Others will be invited to review these determinations in Step 5.

How precise do you need to be?

This process requires a determination of the relative importance of each risk. You want to sort risks into a fitting high-medium-low category: absolute precision is not needed.

Your aim in this step is to have risk determinations that are not wrong. Wrong would be saying "high" and later changing a rating to "low" or vice versa. If the rating is believed to be correct, if you could make a decent argument as to why you rated it that way and you do not expect to be changing it, if you would be open to revision if compelling new information were provided, then that is sufficient precision for this step.

It is recommended that you use available information to rate a risk, as you will likely have many risks that you have to work through and launching a bevy of new research studies will be impractical. Indicate how the determination was made and what information or who was consulted, and move on to the next risk. As new information surfaces and others are later able to provide additional information, you can revisit the determination and adjust it if necessary.

Who should participate?

How you choose to complete your initial risk analysis will be highly influenced by staff capabilities and interests, as well as by the size, subject fields and availability of the expert pool you turn to. Three main types of people should be invited to participate in this step:

Your in-house staff: No one knows your program as well as the staff. If they have the ability to make a first determination about the risks, then you should let the staff do as much as they can. Working on this will also raise their knowledge about climate impacts.

Experts: You probably already turn to specialists or professionals whenever you need to go beyond your in-house knowledge or skills. Experts in climate, environmental sciences or natural resources from government agencies, universities, NGOs or consulting companies who are familiar with your study area could assist your staff with this climate change risk analysis. Experts could also be asked to do the analysis themselves. Local experts (like the director of a public works department or professional landscapers) can also contribute. They might in fact be the best people to estimate the consequence of a risk.

Partners: When you did your initial stakeholder consultation, people may have indicated that they had expertise or knowledge in particular domains. It will be useful to bring those experts back into the process now, especially if they said they could help with this step. For example, it may be useful to have the president of the fishing club weigh in on the consequences of warmer water.

You can approach different people about likelihood and consequence ratings for the same risk. For example, you might ask climate experts about the likelihood of a risk and turn to in-house staff for a determination of its consequences or its spatial extent.

If you turn to experts or selected stakeholders

Carefully consider which risks you would like an expert or selected stakeholders to weigh in on and whether you would like them to comment on the likelihood or consequence (or both) of that risk. Keep in mind that the goal here is to generate an initial analysis. Consult with those who can fill in gaps with particular, needed expertise. In Step 5 you will share what you have done more broadly to help reach agreement. In this step, you only should do what is necessary to get off to a good start.

If you reach out to others in this step, it will be helpful to communicate the following information:

- An overview of the vulnerability assessment process, including identification of your organization's goals.
- Which risks you would like their input on and whether you would like their determination on the likelihood or consequence (or both) of that risk.
- That a qualitative scale is being used to determine the likelihood and consequence for each risk—see "What you (or those you consult) need to determine" below.
- The level of effort that is or is not needed right now.
- That a justification, rationale or source for their input would be helpful.

What you (or those you consult) need to determine

Risks will be ranked on a **qualitative** scale for consequence, likelihood, spatial scale of the impact, and time horizon. It is helpful to note the justification or source used to determine the likelihood and consequence of each risk, as it will be a useful reference later on in the process (if you need to revisit this, or to find out more when you are planning adaptation actions).

The Workbook uses scales with three divisions (high-medium-low). In order for this exercise to be most useful, you must have a spread of likelihoods and consequences for the risks identified. If everything is ranked as high-likelihood and high-consequence, the vulnerability assessment process is going to be less effective as a decision-making tool. In the situation where a disproportionate number of risks are falling into the "high" category, you should probably expand your definition of "medium" to capture more risks.

Using a qualitative scale

A qualitative scale at this stage is useful for several reasons:

- Future changes in the climate system cannot be projected with the exactness that is needed to precisely quantify the probability of a risk at any given future time.
- You might need to contend with well over a hundred distinct risks. It would be prohibitively expensive to scientifically quantify the likelihood of each one, as well as to estimate the cost of damages each risk poses. The cost of doing the risk assessment should not exceed the cost of mitigating risks themselves.
- The general public has a poor ability to process statistical probability. Research on decision theory shows that giving people mathematically identical choices elicits different responses depending on how a problem is framed.
- It will be much easier to reach agreement on a qualitative rank (high-medium-low) than on whether mathematical calculations of likelihood or consequence were done correctly.

Some people might elect to use numbers to express a qualitative value (e.g., rank 1–5). Keep in mind that these numbers are probably symbolizing subjective values because there is no easy mathematical way to accurately derive one true quantitative value for each risk. If you are using numerals to represent opinions, they are still categories. Unless going from category "1" to "2" truly represents the same quantitative change as going from category "3" to "4," and respectively likewise for all the other numerical relationships, you cannot properly add, multiply or divide with the numerals, or find their mean (although median and mode are valid).

Be sure to look ahead to Step 5 and Steps 6–8b to understand how these categories will be used if you are tempted to increase the number of categories or to do math with them. Is it really going to be helpful to use 5-point scales or 10-point scales to generate 25 or 100 different classes of risk? Are you actually going to use that level of differentiation to make decisions? The key thing here is not to overcomplicate the process. Define what you mean by your scale and communicate these definitions to your stakeholders.

It may be helpful to define what each category (high-medium-low) means for your organization.

Note that in the descriptions below for each parameter, the qualitative category "(a)" is always best and "(c)" is the worst.

Consequence

Consequence is the effect the risk would have on your organization's goal were it to occur.

> (a) Low (life will go on; not as important as many other things; could adjust)
>
> (b) Medium
>
> (c) High (major disruption; goal is out of reach or not even attainable)

Likelihood

Likelihood is the chance of the risk actually occurring (i.e., probability). For the risk, you determine how likely it is to affect the goal.

> (a) Low
>
> (b) Medium
>
> (c) High

For planning purposes you can avoid attaching quantitative probability labels to categories. Unless you have access to a large team of top scientists and a lot of supercomputers for simulating the climate at the scale of your study area and running the appropriate hydrological, hydraulic, demographic, land use, and ecological models, your mathematical estimate of probability is not going to be any more accurate than the qualitative category anyway.

Since this is a qualitative analysis—to aid future decision-making by spreading the risks over three categories, you could define your likelihoods so that 20%–40% of your risks fall in each of the high-medium-low groupings (unless you have a large number of risks that are not likely to occur). All the risks in the high category would be more likely to occur than those in the medium category, and likewise for medium and low.

Spatial extent

Spatial extent refers to the proportion of your geographic area that the risk will affect (recall that the Introduction states that this WORKBOOK methodology works best at a spatial scale that is large enough that risks are numerous and diverse and small enough that managers know the territory). Knowing whether problems are isolated or widespread will help you in the action planning process as you decide how best to use limited resources.

> (a) Site (e.g., a few waterfront lots, a bridge, a sewage treatment plant)
>
> (b) Place or region (e.g., community, harbor, state park, wildlife refuge, sub-watershed)
>
> (c) Extensive (most of the watershed or most of the estuary)

If a different spatial scale is more appropriate for your situation, then feel free to modify these categories.

Time horizon

Time horizon until the problem begins will help with developing your action plan. Decisions about low or medium problems that will not emerge for decades could be postponed. Conversely, high-impact problems that are already occurring need some attention right away.

    (a) More than 30 years

    (b) 10–30 years

    (c) Already occurring or 0–10 years

Treat the time horizon for the risk as being independent of the likelihood of it occurring. If you think that absolute dates are more appropriate than relative time periods, choose whatever works best for your situation.

Habitat type

Habitat type is a category that is not necessarily a risk dimension. Identifying it now as you go through your risk analysis and develop your table will become useful later when you work on an action plan in Steps 6–10. When deciding on adaptation actions, you may want to group all of the risks associated with a certain type of habitat. You might, for example, want to see all of the risks associated with tidal wetlands or with residential areas. Looking across all the associated risks can help you choose adaptation actions that address more than one risk.

Resources that can help

Resources about each of the seven types of climate change stressors are listed in Appendix E. These sources could help you to determine how much change is expected for each type. If you are aware of local or regional climate change impact studies from academic institutions, governments or nonprofits, they can also help inform this step.

How to proceed when climate projections give a range of answers

Environmental managers have always had to deal with uncertainty when making decisions. It is important to be transparent about the magnitude of the uncertainty, the range of possible outcomes, and the resulting justification for making a specific decision. Nevertheless, sufficient information currently exists to make good determinations about whether a risk is likely to lead to small or large problems.

For the purposes of this analysis, knowing the direction of the change is often enough. For example, you do not need to determine whether sea level will be 1.25 feet, 1.5 feet or 1.75 feet higher 45 years from now. Instead, think about what your system will be like when sea level is 1.25 feet, 1.5 feet or 1.75 feet higher in the coming decades. Similarly, you probably do not need to determine whether it is going to be 4°F or 6°F warmer. Think about what the impacts will be at the low and high ends of your annual temperature range when temperatures are consistently warmer than they are now, or phenomena happen earlier in the spring or later in the fall.

Urgent problems

Ultimately, if the scale or urgency of a problem is so big that planning estimates are not adequate, then you have to turn to other risk management techniques for those risks. The methodology of this WORKBOOK is not the right approach for problems that need precise, quantified answers.

See Appendix C.

As you are assessing your risks, it may be helpful to indicate the level of certainty associated with each risk. When you evaluate the risks in Step 5—Risk Evaluation: Comparing Risks and later decide which ones to move forward with in your action plan, you may find that there are some risks for which there is great uncertainty and you need more information before investing resources to mitigate them.

When using the WORKBOOK methodology to conduct a vulnerability assessment, use the best available information and use an iterative process that allows you to reassess the identified risks as new information arises.

How to proceed when climate projections give opposing answers

As you move through your risk analysis, you may discover a situation in which climate projections indicate different directions of change. For example, while examining the likelihood of more frequent drought for your watershed, you might discover that some climate models show there will be more precipitation events and less drought, while other models say that drought conditions will become more frequent.

There are lots of climate models, and several standard scenarios that each could run. Because each model is slightly different and the scenarios differ in their assumptions, every combination will have a different result than every other simulation. Sometimes model results cluster in a narrow range; sometimes that range is wider; sometimes the range includes both positive and negative projections of change.

How to proceed is related to how confident you would be about deciding on one direction of change and its approximate magnitude. Your answers to the following questions about model outputs can influence your decision:

- Do you have an outlier model?
- Do you have a continuous range of model results that happens to encompass zero?
- Do you have two clusters of model results?
- Do you have model results that are scattered all over the place?

If your sense is that there is enough of a modeling consensus for you to move forward, then proceed.

But if your sense is that the best available climate change information is saying that anything could happen, then in the process of gathering more information you have now identified a condition that might lead to new risks.

New risks come because your conclusion that climate could change in opposite ways has introduced a new stressor. To continue the drought example, the "increasing drought" stressor would remain and you would now have an opposite "less-frequent drought" stressor too. You might not think that less-frequent drought is a stressor, but it is a change from current conditions and that could lead to some unwanted changes in the environment. The potential for new risks might best be handled by returning to Step 3—Risk Identification and generating a new set of risks that would stem from the opposite stressor. Then come back to this step with any newly identified risks. You will have to decide how to allocate likelihood for the risks associated with the stressor pair. If you concluded that anything could happen then you probably should not assign "high" likelihood to risks related to either half of the duo.

Thresholds or tipping points

Tips occur when small actions, events or changes have large effects on future states. Environmental systems that respond to climate changes by crossing thresholds will complicate your risk analysis. When an ecological threshold is reached, things that were changing gradually (if at all) suddenly change abruptly. A low impact suddenly becomes high, or a local problem is suddenly everywhere, or a present/absent switch occurs. Even if you do not know when a tipping point will be reached, you might be able to anticipate that it will be reached. When sea level rise will push salty water upstream to a sensitive freshwater marsh, or when the winter will become a frost-free season, cannot be known precisely. However, anticipating these thresholds will better prepare you in the event that they do occur.

A risk could have a set of determinations about consequence/likelihood/area/timing that apply before the threshold is reached and another for after. One strategy would be to split the risk into two risks: a before-after pair. An alternative is to keep the risk intact and proceed with your risk analysis by answering some questions:

- Do you need to take action anyway? If this is a risk that will warrant attention even before a threshold is reached, how you choose to respond may also affect the onset of the transition. Mitigating a low risk now may head off the change to a high risk later.
- Is the threshold reversible? If it is crossed, can you take actions that would undo the change, or is it a one-way transition? Losing an endangered species is irreversible. Sometimes reversing the change is theoretically possible but unrealistic. Try to stay in the plausible part of the spectrum.
- How long will an effective response take? If you must respond to the risk if you do cross the threshold, is the time to adequately respond short or long?

If the consequences after crossing the threshold are high, and (because of permits, costs or other foreseeable problems) it would take a long time to effectively respond—whether before or after a tipping point is reached—then this is a high risk, even if the time of onset is unknown.

Use existing information

This application of the risk management process is intended to be completed with existing information and resources. New fine-scale climate modeling is not required for the type of qualitative analysis this vulnerability assessment supports, but use it if it already exists and is easy to obtain.

At the time of this publication, climate model outputs for monthly temperature and precipitation are easily available for U.S. counties or finer resolutions.[3] These should be more than sufficient for a planning-level vulnerability assessment where you are looking to sort risks into high-medium-low categories.

To Get Started

Use the list of risks generated in Table 3-3 and note which ones you will be able to analyze internally and those for which you will want outside expertise.

Assign someone to be responsible for the initial assessment of the four elements of each risk (consequence, likelihood, spatial extent of impact, time horizon). Then ensure that the responsible people provide you with enough information to fill out Table 4-1.

---

[3] EPA's National Stormwater Calculator is a desktop application that estimates the annual amount of rainwater and frequency of runoff from a specific site anywhere in the United States and includes options for historical weather and climate change scenarios. http://www.epa.gov/nrmrl/wswrd/wq/models/swc/

The NEX-DCP30 Viewer allows the user to visualize projected climate change for any county in the continental United States. http://www.usgs.gov/climate_landuse/clu_rd/nex-dcp30.asp

The USGS Derived Downscaled Climate Projection Portal allows visualization and downloading of future climate projections from a group of "statistically downscaled" global climate models. http://cida.usgs.gov/climate/derivative/

TABLE 4-1. RISK ANALYSIS

Organizational goal	Climate stressor	Risk	Is this an opportunity instead of a risk?	Consequence (a–c)	Likelihood (a–c)	Spatial extent of impact (a–c)	Time horizon (a–c)	Habitat type	How confident are you in your risk analysis? Do you have sources that support your determinations?
1.									
2.									
3.									
4.									
5.									
n.									

Consequence:

(a) Low (life will go on; not as important as many other things; could adjust)

(b) Medium

(c) High (major disruption; goal is out of reach or not even attainable)

Likelihood:

(a) Low

(b) Medium

(c) High

Spatial extent of impact:

(a) Site (e.g., a few waterfront lots, a bridge, a sewage treatment plant)

(b) Place or region (e.g., community, harbor, state park, wildlife refuge, sub-watershed)

(c) Extensive (most of the watershed or most of the estuary)

Time horizon:

(a) More than 30 years

(b) 10–30 years

(c) Already occurring or 0–10 years

Additional Resources

Also see Appendix B.

EPA resources about assessing climate change risks

EPA. 2011. *Healthy Watershed Initiative: National Framework and Action Plan*.
http://water.epa.gov/polwaste/nps/watershed/upload/hwi_action_plan.pdf

EPA Region 9 and California Department of Water Resources. 2011. *Climate Change Handbook for Regional Water Planning*.
http://www.water.ca.gov/climatechange/CCHandbook.cfm

EPA. 2007. *National Land Cover Data Classification Schemes (Level II)*.
http://www.epa.gov/mrlc/classification.html

Climate Ready Water Utilities. *Climate Resilience Evaluation and Awareness Tool (CREAT)*.
http://water.epa.gov/infrastructure/watersecurity/climate/creat.cfm

Climate Ready Water Utilities. *Climate Ready Water Utilities Toolbox*.
http://www.epa.gov/safewater/watersecurity/climate/toolbox.html

How sensitivity, exposure and adaptive capacity influence vulnerability

National Wildlife Federation. 2011. *Scanning the Conservation Horizon: A Guide to Climate Change Vulnerability Assessment*. Chapter 3.
http://www.nwf.org/~/media/PDFs/Global-Warming/Climate-Smart-Conservation/NWFScanningtheConservationHorizonFINAL92311.ashx

Considerations for assessing vulnerability

ICLEI. 2007. *Preparing for Climate Change: A Guidebook for Local, Regional, and State Governments*. Chapter 8, pp. 69–71.
http://www.icleiusa.org/action-center/planning/adaptation-guidebook

Climate projections

See Appendix E for sources of climate projections.

USDA. 2012. *Climate Projections FAQ*.
http://www.fs.fed.us/rm/pubs/rmrs_gtr277.pdf

U.S. Geological Survey

What Is "Risk Evaluation: Comparing Risks"?

This portion of risk evaluation is an opportunity for you to reach agreement about the assessment of risks that your organization is facing. After the vulnerability assessment is complete you will pick up the rest of risk evaluation in Step 7—Risk Evaluation: Deciding on a Course of the action planning process.

Objective of This Step

The objective of this step is to develop a consequence/probability matrix and review it with stakeholder input. After you have agreement about your risk assessment, you will have the opportunity to further evaluate your vulnerabilities by looking at goals and habitat types.

Process

After the people you consulted in Step 4 provide you with an initial analysis of your risks, you have an opportunity to turn to a wider set of people. Your partners and stakeholders can react to the determinations about consequence and likelihood, and help you to make sure they are accurate.

Create a consequence/probability matrix

A C/P matrix is a useful communication tool and will help guide decision-making about which risks your organization will address in your action plan. In Step 4—Risk Analysis you determined an initial consequence and likelihood for each risk. In this step you will map each risk to a cell or box on a matrix using those characteristics. You will create a matrix like the example in Figure 5-1.

Ask your stakeholders and advisors to help

Once you have your C/P matrix, reach out to the key stakeholders and partners for their input or concurrence in order to verify that you were on the right track in previous steps. You can reach out to individuals separately or convene a workshop. You can ask some people to just look at some of the risks. There are many ways to get feedback on your initial risk analysis. The more knowledgeable people who can review your C/P matrix, the better (up to the point of diminishing returns). You may find it useful to mark up your C/P matrix (which could be done separately for each goal, stressor or habitat type, or which may span several sheets of paper if printed) to include information from stakeholder consultation. You could also track input with additional notes to Table 4-1.

> ### Vulnerability assessment
>
> After this step, the matrix will be an agreed-on categorization of all foreseeable climate-change-related risks based on their likelihood of occurrence and consequence to your organization's goals. The matrix does not prescribe or mandate any further activities. How or whether your organization intends to address these risks will be considered during the action planning steps that follow the completion of your vulnerability assessment.

There are advantages to asking others to participate. However, do this prudently: the primary purpose here is to get the best risk evaluation you can get. You do not want this to turn into a vote about what is important, nor do you want people angling for their issues to get ranked higher. The vulnerability assessment is about risks to your organization's goals. Try to keep that the focus.

Ask for thoughts on your risk analysis:

- Do they generally agree with how the risks are placed in the C/P matrix?
- Are there additional risks (pertaining to your organization's goals) that should be added to the matrix?

Ask for feedback about the risks themselves:

- Are any of them priority risks for your stakeholders? Alternatively, are there any that they do not want to be addressed?
- Are there stakeholders or partners who are working to address any of the risks through their own programs?

FIGURE 5-1. An example consequence/probability matrix.

		Low	Medium	High
Likelihood (probability) of occurrence	**High**	1. Warmer water may stress immobile biota 2. Warmer water may lead to changes in drinking water treatment processes n. _____	1. Warmer water may hold less dissolved oxygen 2. Sea level rise may cause bulkheads, sea walls and revetments to become more widely adopted n. _____	1. Shoreline erosion from sea level rise may lead to loss of beaches, wetlands and salt marshes 2. Combined sewer overflows may increase from more intense precipitation n. _____
	Medium	1. Increased wildfires from warmer summers may lead to soil erosion 2. Warmer winters may lead species that once migrated through to stop and stay n. _____	1. Parasites and bacteria may have greater abundance, survival or transmission due to warmer water 2. Warmer summers may drive greater water demand n. _____	1. More frequent drought may diminish freshwater flow in streams 2. More intense precipitation may cause more flooding n. _____
	Low	1. Warmer water may lead open seasons and fish to be misaligned 2. Warmer winters may lead to more freeze/thaw cycles that impact water infrastructure n. _____	1. Warmer water may lead jellyfish to be more common 2. Ocean acidification may cause the recreational shellfish harvest to be lost n. _____	1. Contaminated sites may flood from sea level rise 2. Warmer water may promote invasive species n. _____

Consequence of impact

Color key: | Green | Yellow | Red |

A C/P matrix is a tool for visualizing how risks were categorized when they were analyzed for their consequences and likelihoods. Risks plotted in red boxes (upper right), yellow boxes (diagonal from top left to bottom right) or green boxes (lower left) of the matrix are informally referred to as "red risks," "yellow risks" or "green risks" in the remaining steps of the WORKBOOK. The 18 risks shown here and their high-medium-low rankings are used solely for illustrative purposes.

Reaching agreement

Each risk can have only one ranking for its likelihood and one ranking for its consequence. If you turn to more than one person (this could be desirable), you should be prepared to receive a range of responses regarding the likelihood and consequence of a risk.

In Step 4—Risk Analysis you asked qualified people to produce answers that were "not wrong." That is the standard here too. You want an answer that you are even more confident is not wrong. You have to consider the range of input you receive from your many stakeholders and decide how to proceed. Perhaps you will have to make a tough call. Revisit what you developed in Step 4 if irreconcilable differences of opinion arise. Ultimately, it is your responsibility to synthesize the feedback into one determination, as this is ultimately your organization's vulnerability assessment. Keep your organization's needs in mind.

No consensus

Why is there disagreement?

Do you have dissension about both the consequence and the likelihood of a risk or just one parameter?

Does the issue involve the uncertainty surrounding projected climate changes?

Is there unequal knowledge about the risk among the disagreeing parties?

Does everyone understand why others think differently?

Possible fixes

Change the wording to something that everyone can agree about. Maybe it is too broad, or maybe one adjective is the consensus breaker.

Split the risk into separate risks that everyone can agree about. A particular risk may have a high likelihood in one place but low likelihood in another part of your study area: divide it into two risks, one for each area. A risk may have different consequences for different species (e.g., ducks vs. herons). Again, turning it into two risks may resolve differences of opinion.

See if disagreement matters. You are deciding whether a risk belongs in one of three qualitative categories (high-medium-low). Even if feelings are strong about high vs. medium or medium vs. low, the ultimate decision on how to categorize the risk may not matter as much. (If the debate is about the high vs. low categories, you should probably revisit this risk more thoroughly.) Would the risk land in the same red-yellow-green category on your matrix regardless of how the disagreement is resolved?

You don't want to knowingly overrate risks—but if you are going to err, then err on the side of ranking the risk higher. If it is ranked higher it will get more consideration in the action planning to come.

- As you investigate it further in the action planning steps, the correct determination should become apparent and you can revise the risk analysis downward if you find that it was overrated. If you underrate the risk, you may never learn that it should have been rated higher.

- If the risk is in active consideration in the action planning steps, you also may find that an action that mitigates another higher risk could mitigate this one too.

Update the matrix

As you confer with your colleagues, stakeholders and consultants, use their input to revise your matrix. Revise the tables you created in Step 4—Risk Analysis as well (be especially sure to keep track of the reasons for any updates you make).

When you have a revised and completed C/P matrix, you have two important results: (1) a broad, risk-based assessment of climate change vulnerability in your system and (2) agreement among management and key stakeholders about how the climate change risks will affect your organization. Congratulations! There is a lot of work to do still, but you are off to a great start.

Use your vulnerability assessment and the C/P matrix as a communication tool. It may be helpful to do the following:

- Publish it online so that it is available within your community.
- Share it within your peer network (e.g., with other National Estuary Programs, National Estuarine Research Reserves, municipalities or watershed associations).
- Start gathering information to fill in data or knowledge gaps.

Get a deeper understanding

You can gain additional insight about your organization's climate change vulnerabilities by organizing your risks into compilations.

Goal: Create a list of all the risks that are related to each of your organization's goals. You could find that some of your goals will be difficult to attain if the climate changes as expected. If, for example, your organization focuses on nonpoint sources of pollution and you find that a number of high risks are going to bear on that goal, you will want to pay attention to that as you adapt your organization to climate change.

Habitat type: In Step 4—Risk Analysis, you had the option to assign a habitat type to your risks. If you did so, you are ready to create a list of the risks in each environmental system. Looking at how much risk is concentrated in, for example, wetlands, urban areas or estuarine waters could help you to design action plans that will lead to sustainable systems.

Keep in mind that these compilations are simply new sortings of information from your risk assessment and risk evaluation. They offer a new angle on your vulnerability assessment that will also be valuable as you move on to develop an action plan to mitigate risks.

To Get Started

Transfer the risks in Table 4-1 (which indicated the ranking for the consequence and likelihood of each risk) to a matrix like the example shown in Figure 5-1.

Additional Resources

Also see Appendix B.

EPA resource about working with stakeholders and coming to agreement

EPA. 2013. *Getting in Step: Engaging Stakeholders in Your Watershed.*
http://cfpub.epa.gov/npstbx/files/stakeholderguide.pdf

Stakeholder participation

NOAA Coastal Services Center. 2007. *Introduction to Stakeholder Participation.*
http://csc.noaa.gov/digitalcoast/sites/default/files/files/1366311008/stakeholder_participation.pdf

NOAA Coastal Services Center. 2010. *Introduction to Planning and Facilitating Effective Meetings.*
http://csc.noaa.gov/digitalcoast/sites/default/files/files/1366310745/planning_and_facilitating_effective_meetings.pdf

Diane Haslem via Partnership for the Delaware Estuary

What Is "Establishing the Context for the Action Plan"?

In this step, you will re-examine the context of your organization to help create and implement an action plan that responds to the risks identified in your vulnerability assessment.

Objective of This Step

The objective of this step is to explore opportunities and constraints that influence your organization's choices, and develop a list of potential partners that could help you in addressing the risks you identified in your vulnerability assessment.

Process

In **Step 2—Establishing the Context for the Vulnerability Assessment**, you pulled together your organization's goals to help you work through the vulnerability assessment. In your vulnerability assessment you identified and assessed all the foreseeable climate change impacts that could affect your ability to meet your goals. This step (**Step 6**) is about your context and the partners you could be working with to implement adaptation actions.

Organizational context

The political, regulatory and cultural environment in which your organization exists influences your ability to take on different projects and accomplish goals. You want to uncover any circumstances that might guide you to choose certain adaptation approaches or prevent you from selecting others. If you choose to mitigate a risk you will need to have the organizational capabilities, resources and commitment to do what needs to be done. If you decide to accept or avoid a risk, you need to know how that will affect not just your goals but your organization too.

Table 6-1 has a checklist to ensure that you review and note the relevant parts of your context. If the review of your situation identifies special implications for any of your risks, you should also note that.

Adaptive capacity

A risk-based vulnerability assessment examines how climate change stressors will affect the ability of an organization to reach its goals. When risks exist, an organization that does not want to accept the possibility of a risk's consequences will have to adapt. An organizational quality known as "adaptive capacity" is used to characterize its ability to make those adjustments.

The adaptive capacity concept can also be used with ecosystems or elements of ecosystems in reference to how much they can cope with climate change. You probably accounted for the adaptive capacity of ecosystems in your vulnerability assessment when you determined the likelihood or consequence of environmental risks.

In reference to organizations, USGCRP's SAP 4.4 report describes four categories of barriers to adaptation (Chapter 9.5) that can constrain your ability to adapt:

- legislation and regulation;
- management policies and procedures;
- human and financial capital; and
- information and science.

The report goes on to suggest some opportunities to overcome those barriers. Keep in mind that to respond to your risks, your first adaptation actions might need to be structural or institutional.

U.S. Climate Change Science Program. 2008. *Preliminary Review of Adaptation Options for Climate-Sensitive Ecosystems and Resources*. Final Report, Synthesis and Assessment Product 4.4.

TABLE 6-1. REVIEW OF YOUR ORGANIZATIONAL CONTEXT

Contextual area	Barriers	Opportunities
Social		
Technical		
Administrative		
Political		
Legal		
Economic		
Environmental		

Partnerships

You undoubtedly have partners who have a similar vision for your watershed. Hopefully, you have been working with them throughout your vulnerability assessment. Partners will be essential in the action planning half of climate change adaptation too. They can help you deal with risks that you may not be able to handle yourself. Solving problems together, dividing up tasks, identifying lead agencies, and finding who can contribute human or financial resources will help everyone to achieve your common goals.

Although all of the risks you identified in your vulnerability assessment could prevent you from achieving your goals, your organization may not be able to mitigate every risk. The availability of partners who can help with a risk may make a big difference in the approach you select. Be aware that some of your contextual limitations may apply to your partners as well.

Use Table 6-2 to note any common organizational goals, objectives or work areas where potential partners could help you. This list does not represent a commitment by either party. The list you develop here is an aid for the next step, where you may run into the limits of what your organization can do alone.

TABLE 6-2. POTENTIAL PARTNERS

Partner/Organization	Common Goal/Objective/Work Area
1.	
2.	
3.	
4.	
5.	
6.	
7.	
n.	

To Get Started

In **Step 1—Communication and Consultation** of the vulnerability assessment, you may have asked stakeholders if they were able to help implement actions to reduce risks. Review your communication plan to see who said yes.

Additional Resources

Also see Appendix B.

EPA resources on building partnerships

EPA. 2008. *Handbook for Developing Watershed Plans to Restore and Protect Our Waters.* Chapter 3.
http://water.epa.gov/polwaste/nps/upload/2008_04_18_NPS_watershed_handbook_ch03.pdf

EPA. 2008. *EPA's Environmental Justice Collaborative Problem-Solving Model.*
http://www.epa.gov/environmentaljustice/resources/publications/grants/cps-manual-12-27-06.pdf

Additional guidelines on creating partnerships

USDA. 2011. *Responding to Climate Change in National Forests: A Guidebook for Developing Adaptation Options.* pp. 22–28.
http://www.treesearch.fs.fed.us/pubs/39884

NOAA. 2010. *Adapting to Climate Change: A Planning Guide for State Coastal Managers.* Chapter 3, pp. 19–23.
http://coastalmanagement.noaa.gov/climate/docs/adaptationguide.pdf

Additional considerations for your context, and the STAPLEE criteria

NOAA. 2010. *Adapting to Climate Change: A Planning Guide for State Coastal Managers.* Chapter 3, pp. 52–54.
http://coastalmanagement.noaa.gov/climate/docs/adaptationguide.pdf

FEMA. 2003. *Developing the Mitigation Plan: Identifying Mitigation Actions and Implementation Strategies.* Chapter 2.
http://www.fema.gov/media-library/assets/documents/4267

Dave Gatley/FEMA News

What Is "Risk Evaluation: Deciding on a Course"?

This part of risk evaluation revisits your vulnerability assessment to take a closer look at your risks to determine which ones your organization will move forward with in the action planning process. Deciding on what actions you want to pursue and whether an action will work are going to be postponed until Step 8a and Step 8b. Here, in Step 7, you will be deciding at a high level how you want to approach your many risks.

Objective of This Step

The objective of this step is to decide whether your organization will mitigate, transfer, accept or avoid each risk.

Process

Although the focus of this step is high-level, the decisions you make here will determine what kinds of adaptation actions your organization implements and which risks you focus on in your action plan, so think carefully about how your organization will approach each risk.

To accomplish the objective of this step requires a little bit of finessing the chicken and egg dilemma. In a systematic methodology, like the one this WORKBOOK uses, when you have scores of risks and limited resources, the dilemma is:

- You can choose to mitigate each risk or use some other risk handling approach for each one. Should you decide your approach before you even know whether there would be any suitable mitigating actions you could take?

- Investigating the options for mitigating a risk will take time and resources. You already know how many risks are in your vulnerability assessment and that you might never be able to mitigate each one; further there will be some that you know you never will act to mitigate. Should you still expend the effort to investigate options for mitigating every risk so that you can make a logical informed choice about whether to mitigate them?

Basically—should you decide what to do and then investigate it further, or when you can't do everything should you still investigate it all before deciding what to do? Trying to have one logical, systematic process is what introduces the procedural difficulty. By definition, there is no easy way to resolve a dilemma, but you will try to make it a smaller problem from two directions:

- First you will get some familiarity with the mitigating actions that are circulating in the planning world. Knowing that options might exist for your situation will help you to decide whether mitigating any risk looks likely to succeed.

- Second, because mitigation requires you to expend resources, you will try to screen out risks that seem suitable for another approach.

You want to filter the set of risks to decide where you will focus your resources. The idea is to head out of this step with a reduced set of risks that you would prefer to mitigate and for which you might be able to supply an appropriate mitigating action. Recording your high-level approach for each risk in Table 7-1 will set you up for Step 8a—Finding Adaptation Actions and Step 8b—Selecting Adaptation Actions, where you will make finer deliberations about how and whether to proceed.

How are risks related?

Your vulnerability assessment has lots of information to help you close in on the course you want to follow for your risks. In Step 5, you concluded the vulnerability assessment by getting a deeper understanding of how your risks are related to each other. Risks may affect a certain habitat type. A set of risks may threaten particular organizational goals.

Mitigating actions you take for some risks might also have the co-benefit of mitigating related risks. Gaining awareness of how risks are related might help you identify a combination of mutually reinforcing risk-reducing actions, which will help you to decide which risk management approach you want to take.

Is mitigation realistic?

Mitigating a risk means acting in a way to modify its likelihood or consequence. Right now you simply want to increase your understanding of the types of actions that would be necessary to lower your risks.

You want to get yourself to the point where you can say that it is realistic to think that a risk could be lowered if you elected to work on it: then you can make a more informed decision about whether you want to mitigate each risk or take a different approach.

The Climate Ready Estuaries *Synthesis of Adaptation Options for Coastal Areas*, as well as other documents (see the "Additional Resources" section of this step and Step 8a), will provide a sense for what might work for your location and combination of risks. You might also recognize that some actions address more than one risk.

While you will not be selecting actions in this step, having some familiarity with ones your organization could take will be helpful. A fair amount of work remains in Step 8a and Step 8b before you will select a specific mitigation technique to use. Do not be too critical right now. In the same way the Workbook advises not to dismiss risks while you are preparing your vulnerability assessment, you should not dismiss actions here right now. You will be tempted to bring in your organization's context or other knowledge to eliminate possible actions. However, judgments about how realistic something is will be coming back into this very soon.

For this part of Step 7, just understand what kind of mitigating actions are in circulation and use your professional knowledge to simply think about whether any would work to lower your risks if you (or someone else) could implement them.

Approaches for each risk

Although these decisions will be made at a high level, you are deciding an approach for risks that you determined have the potential to affect your organization's ability to meet its goals. You need to be prudent as well as aware that how you choose to approach your risks will determine where your adaptation actions are focused going forward.

Your context and constraints will have a big influence on the approach you select. It will be helpful to consult partners and stakeholders as you determine the approach for each risk. Some risks may be very important to a stakeholder group, or work may already be underway to address other risks. It will also be helpful to consider the spatial scale of the impact. You may opt for a different approach depending on whether it affects a small or large area.

Risk management uses four general approaches for responding to any given risk:

- Mitigate.
- Transfer.
- Accept.
- Avoid.

Mitigating a risk is the one approach in which your organization would be making changes to the risk path (Step 3). The other approaches handle risk more administratively.

Mitigate

Mitigating a risk involves taking actions to reduce the likelihood and/or consequence of the threat to your goals.

Transfer

"Transfer" is a technical risk management term for having another organization take responsibility for reducing the risk. Your risk is mitigated by another party. Buying an insurance policy is an example of transferring risk by having another party reduce the consequences if the risk occurs. An environmental management organization will probably not be buying insurance policies, but transfers can occur when other organizations will act in ways to reduce your risks. Maybe your risk could be mitigated when a highway is rebuilt or as part of other infrastructure work. Maybe you can agree to lead a reforestation effort if a partner agrees to restore some other habitat. Maybe some agency has announced it will be taking hazard mitigation actions that would have the co-benefit of reducing some of your risk. If others' actions are lowering the likelihood or consequence of your risks then you have transferred the risk reduction responsibility to them. You aren't mitigating the risk, but it is being reduced. Note that you cannot unilaterally transfer a risk. Other organizations need to affirm that they will actually mitigate the risk; otherwise the risk will still be there. You can opt to transfer some of a risk by partnering with another organization or by making a financial contribution to someone else's mitigation project. If you are working with partner organizations on your adaptation plan, this is an opportunity to decide which organization is going to be the lead for which risks.

Accept

You may have to consider accepting a risk if you are unable to mitigate or transfer it through a viable strategy. Accepting a risk means that your organization will continue with business as usual and run the risk, dealing with the impact if/when it does occur. You might choose to accept a risk for some time, and then later begin to work on mitigating it. Reasons to accept a risk could include a long time horizon before impacts are expected, inability to locate a suitable mitigation strategy, or a lack of worry about the consequence of the risk. If more resources or information become available, you can re-evaluate and decide if there is an opportunity to use another approach besides "accept" for a risk.

If you don't make a choice to use any other approach, then by default you have elected to accept a risk. You are running the risk.

Avoid

The mitigate/transfer/accept approaches assume that you do not want to change your organization's goals. If those three approaches are not feasible for you, then you may want to avoid a risk by narrowing goals that are related to this risk. Avoiding is about being out of the risk altogether so it will not affect you. Typically, avoiding a risk involves eliminating its root cause. However, since your organization by itself will not stop climate change from occurring, and since you cannot relocate your place, avoiding a risk in this context would require a shift in your organization's operations or goals so that you are no longer exposed to that risk. For example, if one of your goals is to maintain some resource, but you conclude that climate change is making that a lost cause, you could relinquish that goal. Avoiding a risk does not mean the impact to the resource or to your place goes away—this is an administrative handling of risk in which you move away from this objective and you will no longer put resources toward it. Avoiding a risk may be a radical move for an organization: you will be pulling back from work that you thought was important. Reserve this approach for risks that will not be mitigated (by you or others) and where accepting the risk is a bad proposition.

Avoid vs. accept

The WORKBOOK often states that if more information becomes available you should go back to earlier steps. In the iterative process of risk management, choosing "avoid" for a risk appears to be something new that has lots of other implications for your adaptation plan. If you are renouncing a goal, then all the risks you identified in Step 3, which derive from that goal, would seem to vanish too. Thus those risks would fall out of your matrix and vulnerability assessment. If they are not in your vulnerability assessment, they would not make it into this step of action planning either. Logically, deciding to avoid a risk makes it and everything related to it disappear.

Instead of acting right away on this logic, you should let your choice of "avoid" sit for now. The implications to your risks and goals arising from an iteration do not go away but are not an immediate problem. Instead, you want to leave open the possibility that even more information could be coming. The risk may turn out to be more benign than you thought; a new risk mitigation technology might emerge; a new partner may agree to help you with the risk; windfall funding might come in; etc. If you expunge the risk and the goal from your vulnerability assessment and action plan, you have nothing to go back to if you do get more information. Keep the risk and the goal, and keep track of the fact that you chose to avoid the risk. Things might change that would eventually let you select a mitigate/transfer/accept approach.

Tracking a risk you choose to avoid keeps a goal on the books. You stop working on the goal itself but do not drop it completely. Pragmatically, it really does not matter what your goal is or is not if you are not going to put any resources into trying to reach it. The difference between the "accept" and "avoid" approaches to a climate change **risk** thus become the difference between "business as usual" toward a **goal** and "no action" toward a **goal**.

Risk management approach	Description	How your organization would use this approach
Mitigate	Take action to lower the consequence or likelihood of the risk (or both).	Address the risk, or lead the effort to address the risk.
Transfer	Another party has responsibility for mitigating the risk.	Allow or ask others to take the lead; assist as you can.
Accept	Run the risk. Accept that the consequences may occur.	Business as usual in spite of the risk. Monitor, and reassess options in the future.
Avoid	Take organizational or administrative action so that you will not be exposed to the risk.	Stop putting resources toward the goal that would be affected. Or delete/revise your goal and thus be out of the risk altogether.

FIGURE 7-1. Examples of how different risk management approaches would be applied.

For this scenario…

Goal: control point sources of pollution.

Climate change stressor: more intense precipitation.

Risk: combined sewer overflows may increase.

…choosing different approaches for handling the risk could set your organization on differing courses:

Mitigate: Your study area is already experiencing increased precipitation, so you will look for specific actions that you could take to reduce the likelihood of more intense precipitation causing further CSO problems.

Transfer: Another entity in your study area is already working to reduce CSOs. Your organization will participate in this effort but will not be the lead.

Accept: Your organization is going to accept that CSOs may increase from more intense precipitation. No action is identified at this time, but this risk will be monitored and reviewed.

Avoid: Your organization recognizes that it is not effective to continue to work on controlling point sources of pollution given that CSOs are going to increase from more intense precipitation. Resources will be realigned to focus on other goals.

Choosing an approach for each risk

Every risk that is on your matrix—whether in the high, medium or low category—represents a potential event that you determined could keep your organization from meeting its goals. In many cases you can transfer all or some of the risk to a partner, you can simply accept and run a risk, or you can change your goals so you avoid it altogether. If these are feasible or palatable then they are what you should do.

What you cannot transfer and do not want to accept or avoid, you must mitigate yourself.

Mitigating a risk: Every risk that has a potential action that could lower it could be assigned to this category. Actions that are easy, that you would do in pursuit of no-regrets or win-win solutions, or that you simply think your organization should do, should be in this category.

Accepting a risk: If you turn to your vulnerability assessment, you will find a time horizon determination for each risk. All green risks in your C/P matrix (Step 5) are good candidates for an accept approach. The farther out in time they seem to be, the better the accept approach gets. Yellow risks that are decades away are also good candidates for the accept approach.

Not accepting a risk: Risks with a high impact (red risks in your C/P matrix) are bad candidates for the accept approach. You and your team identified these risks as highly likely to derail your organization. In order to continue to function as an effective organization that can meet its goals, you should probably choose another approach for these high-impact risks.

Risks you could accept but should assign for mitigation: These are the green risks that are happening now as well as the yellow risks that are expected within the next 10–30 years. The green risks are low-consequence problems anyway, and you have some time to respond to the yellow risks (unless the response requires a very long lead time). Assigning them to the mitigation approach has the potential advantage of keeping them moving through your adaptation planning steps.

If you assign these green or yellow risks to the accept approach, they will be out of sight and out of mind as you create your action plan. The advantage of bringing them forward is that they will be in front of you when you are trying to decide the mitigating actions you do need to pursue. Actions that might address higher risks could address these lower risks too. You might select some actions over others if you are actively aware that they address lower risks too. Carrying these risks into the mitigation planning (Figure 7-2) may lead to better organizational choices about actions than if the risks are assigned to the accept approach.

FIGURE 7-2. Some risks that are logical candidates for the "accept" approach could be strategically assigned to the "mitigate" approach based on their time horizons.

	Green risks	Yellow risks	Red risks
(a) Risks that are logical candidates for the "accept" approach  *None* *Risks that are more than 10 years away* *All*	All, i.e., time horizon: • Already occurring or 0–10 years • 10–30 years • More than 30 years	Time horizon: • 10–30 years • More than 30 years	None
(b) Risks that are logical candidates for "accept" but could be strategically assigned to the "mitigate" approach 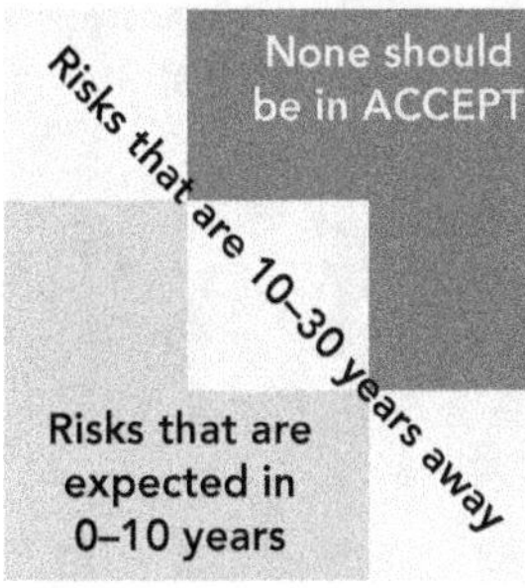 *None should be in ACCEPT* *Risks that are 10–30 years away* *Risks that are expected in 0–10 years*	Time horizon: • Already occurring or 0–10 years	Time horizon: • 10–30 years	Not applicable
(c) Risks that would remain assigned to the "accept" approach  *None should be in ACCEPT* *Risks that are more than 30 years away* *Risks that are more than 10 years away*	Time horizon: • 10–30 years • More than 30 years	Time horizon: • More than 30 years	None

Color key: Green | Yellow | Red

Further, if the climate changes faster than you expect or if you were too optimistic in your time horizon determination, then by the time you complete another iteration of your vulnerability assessment, these unwanted risks may already be upon you.

If you are not sure about mitigation: You are not obligated to start responding, but if you think you would want to mitigate a risk in the future, then you could benefit from the risk management process now by assigning it to the mitigation approach. In the systematic risk management process, only those risks that you identify for mitigation will be carried forward to Step 8a and Step 8b, where you will select adaptation actions for the risks, and to Step 9, where you will develop your plan. The other risks—which you are deciding not to actively mitigate but to transfer/accept/avoid—will be picked up in Step 10—Monitoring and Review after you write your plan. If Step 8a and Step 8b show you that there is not a viable mitigation strategy for a particular risk, you can return to this step and choose another approach.

Approving the approach

After you have investigated and chosen a high-level approach (mitigate/transfer/accept/avoid) for every one of your risks, your organization's key decision-makers (board of directors, management committee, staff, and the people who are regularly involved with deciding what you do) should agree before you go further with your action plan. Your organization is preparing a major strategy document. While this is not a new strategic plan, it is a statement of how you plan to continue being able to achieve your mission and goals.

If you selected the avoid approach for any of your risks, you need to carefully think through what this means. You have chosen to withdraw from whatever goal would be thwarted by the risk. This may be the whole goal if the goal is narrow (e.g., maintain critical habitat for a particular bird species), or it may be a part of a goal if the goal is broad (e.g., maintain ecological diversity in your watershed). If you are deciding to withdraw from pursuing a goal (and thereby in effect rewriting your strategic plan), key decision-makers should approve this change in course.

FIGURE 7-3. Your organization can opt for any risk management approach that serves its needs. The flow chart depicts a logical sequence that could help with decision-making when resources are limited and not every risk can be mitigated.

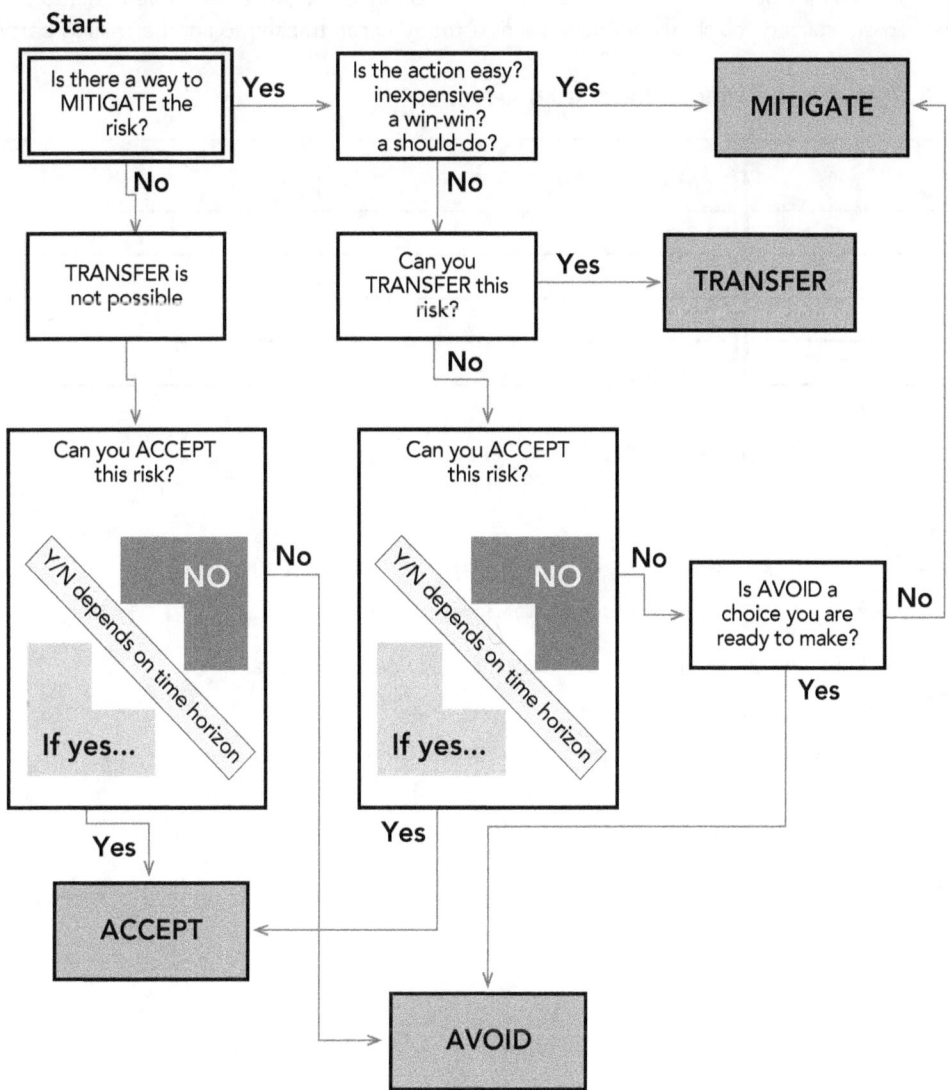

As you consider each risk, the first question is whether there is even the possibility of mitigating it. Did you come across any action that you or others could take to change the risk path so that the risk's consequence or likelihood will be reduced? If no (there is not a feasible way for you or anyone else to lower a risk to a tolerable level), then you either have to accept it or avoid it. If yes (there is a way to mitigate the risk), then mitigation could be your first choice if it is a low-intensity action or something you want to take on. Transferring all or some of the risk for someone else to mitigate could be the next best choice. If no one else will agree to the risk transfer, then you could accept the risk if it is low-impact, or medium-impact with a longer time horizon (see text earlier in this step for a discussion about selecting the accept approach). Otherwise, if you are not prepared to avoid the risk, you have to mitigate it yourself.

To Get Started

You need to mitigate everything that you cannot transfer and do not want to accept or avoid (Figure 7-3). Deciding which risks you want to mitigate may not be the best way to begin. For a first round of decision-making it may be easier to leave the mitigation option for last, after you have rejected the other strategies. To get started you should identify the risks that you can transfer to another willing party.

TABLE 7-1. RISK IDENTIFICATION: CHOOSING AN APPROACH

Risk	Is the risk in the red, yellow or green zone of your C/P matrix?	Approach (mitigate/transfer/accept/avoid)
1.		
2.		
3.		
n.		

Additional Resources

Also see Appendix B.

EPA resources for climate change adaptation options

Appendix F—Actions That Could Reduce Water Temperature

Climate Ready Estuaries. 2009. *Synthesis of Adaptation Options for Coastal Areas.*
http://www.epa.gov/cre/

Climate Ready Water Utilities. 2012. *Adaptation Strategies Guide for Water Utilities.*
http://water.epa.gov/infrastructure/watersecurity/climate/upload/epa817k11003.pdf

Approaches for risks

National Research Council. 2010. *America's Climate Choices: Adapting to the Impacts of Climate Change.* pp. 90–120.
http://nas-sites.org/americasclimatechoices/sample-page/panel-reports/panel-on-adapting-to-the-impacts-of-climate-change/

Project Management Institute. 2008. *The Standard for Program Management.* 2nd ed. 11.4.2.

National Wildlife Federation. 2011. *Scanning the Conservation Horizon: A Guide to Climate Change Vulnerability Assessment.* pp. 9–11.
http://www.nwf.org/~/media/PDFs/Global-Warming/Climate-Smart-Conservation/NWFScanningtheConservationHorizonFINAL92311.ashx

Partnership for the Delaware Estuary

What Is "Finding Adaptation Actions"?

In this step you will be taking a look at the risks you are choosing to mitigate and choosing a set of adaptation actions that could be effective. After this step the list will move forward for more evaluation in planning-level Step 8b and more investigation at the project level when you kick off work in Step 9.

Objective of This Step

The objective of this step is to develop a list of adaptation actions that your organization wants to further assess before deciding to implement them.

Process

You will identify potential adaptation actions for the risks you selected for mitigation in Step 7. At the end of this step you will have a list (Table 8a-2) of adaptation actions that address those risks. Recall that the risks you decided to transfer, accept or avoid will be picked back up in Step 10—Monitoring and Review, since your organization is not going to take any actions on them at this time.

Recapping what has come before to help with what is next

In Step 3 you identified each risk by describing how a climate change stressor might affect your organization's goals. Stressors and consequences are built in for every risk.

In Step 4 and Step 5 you reached agreement about the consequence and likelihood of each risk and mapped each risk to the corresponding location in a C/P matrix. The C/P matrix is divided into three impact zones: high (red), medium (yellow) and low (green); each risk is within one of the impact zones (in shorthand, each risk is either red, yellow or green).

In Step 7 you may have familiarized yourself with a range of adaptation actions (such as in the CRE *Synthesis of Adaptation Options*) to help you with decisions about whether to mitigate/transfer/accept/ avoid the various risks.

In Step 7 you selected which risks your organization wants to mitigate. In this step and Step 8b you will find and select the mitigating actions your organization will pursue.

Adaptation actions

For every risk you know that the **no-action alternative** is this: when time catches up with you or your luck runs out, then the consequence hits. You already decided in Step 7 that there are a number of risks that you want to mitigate because you did not want to run them (i.e., did not want to accept them). Thus you already determined that you want to keep those no-action scenarios from occurring.

In Step 7 the WORKBOOK says: "Mitigating a risk involves taking actions to reduce the likelihood and/or consequence of the risk on your goals." Thus, the **purpose** of an action is to reduce the likelihood or consequence of a risk—ideally substantially reducing the risk. You mitigate risks by implementing adaptation actions that move risks toward the lower left (green) corner of your C/P matrix (Figure 8a-1).

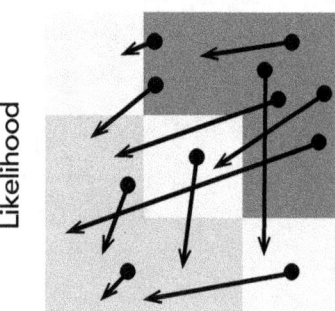

Consequence

FIGURE 8A-1. Risks are mitigated by taking actions that would lead them to be replotted closer to the lower left corner of your C/P matrix. If you reduce the consequence or the likelihood (or both), you have reduced the amount of risk.

Breaking the chain

A mitigating action works somewhere in a risk's cause–effect path to make a difference in the severity of the consequence or in the likelihood of it ever happening (Figure 8a-2).

FIGURE 8A-2. Risk mitigation has to happen in the "business as usual" part of this risk path diagram from Step 3. To change the likelihood or consequence of a climate change risk (and make the goal achievable), business as usual (either the way the environment functions or the projects the organization chooses to execute) needs to be different.

To mitigate a risk you need to find actions that change course from the business as usual plan so that your goal can still be realized. You need to understand what happens in the "business as usual" cloud in Figure 8a-2, and you need to rely on your professional knowledge to conceive how an action could produce a different outcome.

Finding where you could act

If your risks follow the format described in Step 3 then you already have a lot of information about the risk path:

> Stressor X could _____, and the result is that we might not attain Goal Y.

The "blank" in this template is what you need to alter so that you achieve Goal Y regardless of Stressor X. You need to understand the unwanted cause–effect path so that you can find an action that will disrupt it.

As you search for meaningful intervention points, you can imagine the risk path in your head or you can use a graphical illustration to help you understand the system. In Step 3—Risk Identification you may have used a graphical description of your system (like the U.S. Postal Service diagram in Figure 3-2 or a conceptual diagram as described in Appendix D) to help identify how stressors could act to produce risks. You can use similar diagrams or turn to the same ones you used in Step 3.

The more detailed your understanding is, the more opportunities you can see for where a mitigating action would have some potential. Figure 8a-3 shows how to use a conceptual diagram to trace a risk path and aid in finding good intervention points. Diagrams like these do not show you what to do; they show you where you could act to disrupt a risk's cause–effect pathway. Next you find actions that could act at those spots.

Knowledge gaps

You may have a high risk that you must mitigate, but still have a poor understanding of how its environmental system functions. A research project may need to be the first mitigating action. But don't let a desire for perfect understanding become an excuse for inaction.

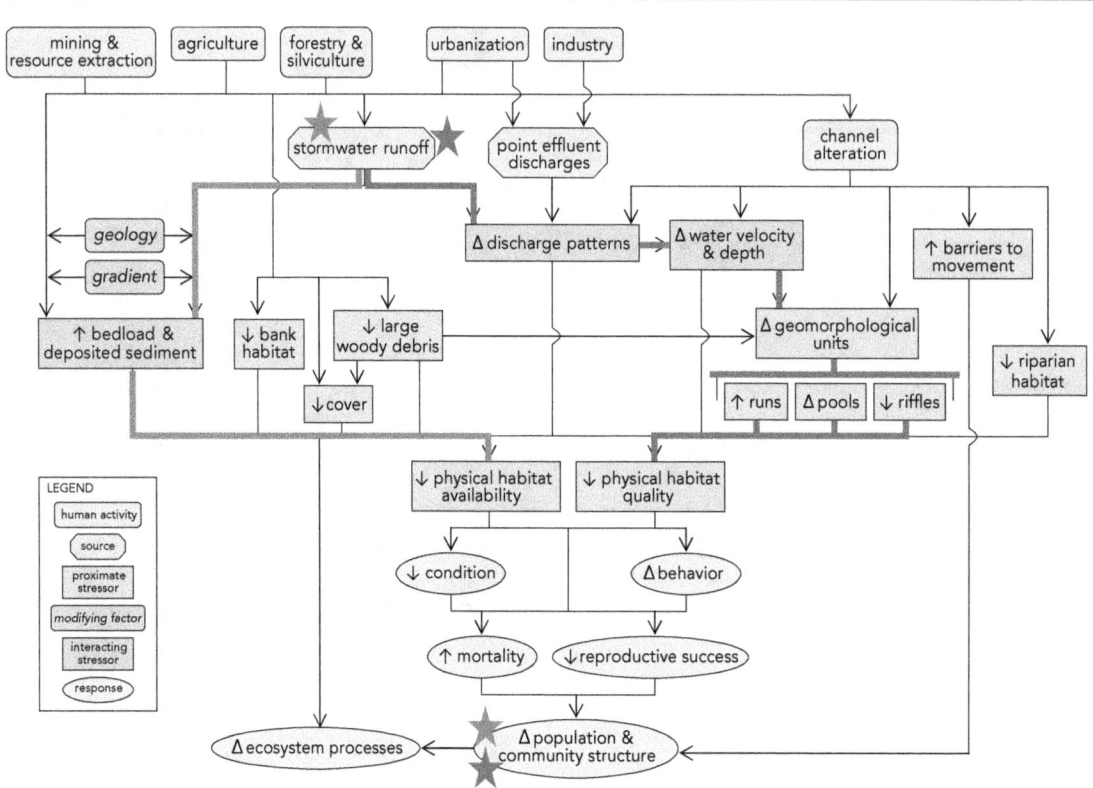

Figure from EPA's CADDIS (see Appendix D)

FIGURE 8A-3. A conceptual diagram with detail of how a system functions will show many places where you could intervene to disrupt the risk path. It might also show how actions that alter other model paths could lower impacts from climate change too.

In this diagram, chains connect stormwater runoff (starred) and the population and community structure of various taxa (starred). You can see how increased precipitation from climate change can work through the system. Stormwater runoff drives two response paths, one (red) affecting habitat through effects on sediment and another (blue) related to how the stream responds to the volume of water. Any action in either path that keeps increased precipitation from having an unwanted ecological effect would change the likelihood or consequence of that climate change stressor producing an unwanted impact.

Secondly, seeing more of the system than the pathways from stormwater to the biological population reveals more opportunities to intervene. If the volume of stormwater seems to be increasing, you can look at controlling runoff or discharge from other land uses. You could look at establishing other riparian habitat to compensate for the climate impacts. You could remove barriers to movement so populations could find refuge during extreme events.

The fuller your understanding of your system, the more places you can find where actions could lower the risk. You want as many options as you can find because some will be better than others.

Finding mitigating actions that look promising

As you find the points along a risk pathway where opportunities exist to mitigate the risk (Figure 8a-3), you need to rely on your knowledge about your system to think about what actions might be promising. You are looking for actions to apply at the relevant spots in the risk path in order to lower the likelihood or consequence of a risk.

Your professional knowledge and training may lead you in certain directions. As you were familiarizing yourself with options in Step 7, additional ideas may have interested you. If your organization goals are similar to the Clean Water Act goals, then Table 8a-1 of the WORKBOOK presents the actions from CRE's *Synthesis of Adaptation Options for Coastal Areas* in a new format that you might find useful. If your organization has goals specific to the management of water utilities, see resources from EPA's Climate Ready Water Utilities program. For other sets of organizational goals, the "Additional Resources" section at the end of this step points to sources of potential actions.

There are hundreds of potential actions that would alter a risk path and reduce the risk's consequence/probability. Many actions are trivial; some are practically impossible. You want to identify a **plausible range** of actions in this step. This is probably not as open-ended a brainstorming exercise as it sounds, but do not reject things that are novel. You want a range of actions because in Step 8b you may decide that some are unpalatable for various reasons and should be rejected. Having a plausible range of actions will help you find opportunities to use actions that might mitigate more than one risk, bring co-benefits, or open up other opportunities. The best way to solve one particular risk might not be as advantageous as a second-best way that mitigates other risks too. It is fine (probably desirable) to identify the same action for more than one risk.

Using Risk Pathways to Find Adaptation Actions

In two Climate Ready Estuaries projects, the San Francisco Bay Estuary Partnership and the Massachusetts Bay Program each worked with EPA's Office of Research and Development in intensive collaborations to identify climate change risks. Parts of the SFEP project are used here to illustrate how conceptual models can be used to find adaptation actions, and to show how some actions address multiple problems.

A number of SFEP's management goals are related to salt marshes:

- Restore healthy estuarine habitat to the Bay-Delta, taking into consideration all beneficial uses of Bay-Delta resources.

- Stem and reverse the decline of estuarine plants, fish and wildlife, and the habitats on which they depend.

- Ensure the survival and recovery of listed and candidate threatened and endangered species, as well as special status species.

- Protect and manage existing wetlands.

- Restore and enhance the ecological productivity and habitat values of wetlands.

Sea level rise and altered hydrology are leading to increased inundation, changes in water quantity and quality, and patterns of sedimentation and erosion, which are interacting with other human activities to create risks.

As part of this project, a more expansive model of salt marsh processes was narrowed (to what is shown) to focus on the factors affecting marsh sediment retention. Experts reached agreement about which relationships had the highest impact under given climate change scenarios. Then they used the diagram to isolate top pathways (pathways are colored blue, green and purple in Figure ES-2 of the original report).

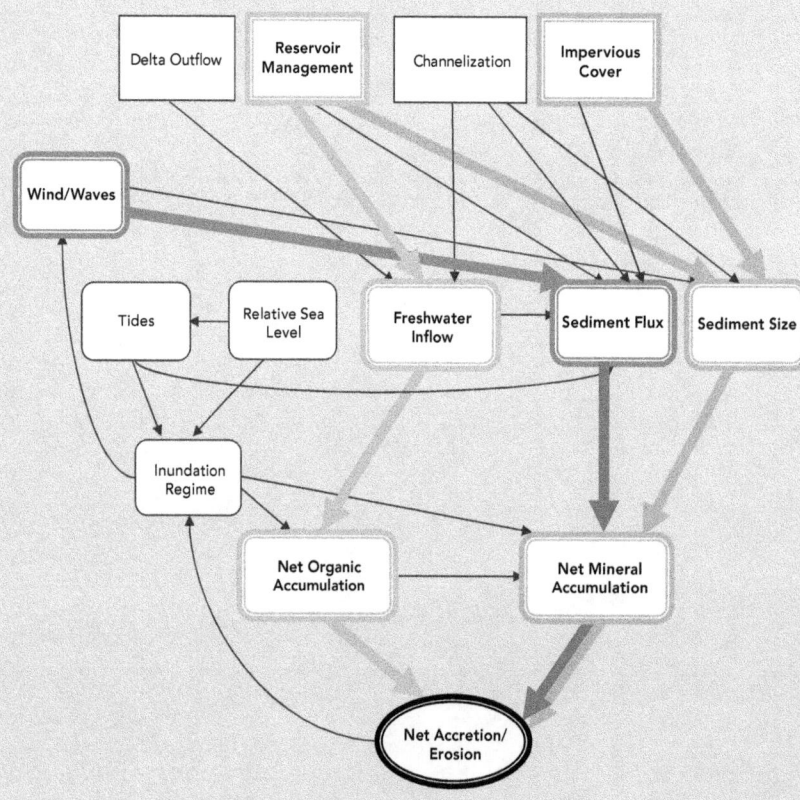

With the three top pathways graphically highlighted, opportunities emerged to find risk mitigation options. While keeping SFEP's management goals in mind, participants used their judgment to identify actions that would respond to the climate change stressors. Understanding how the system operates pointed to a dozen strategies that could be used to help promote accretion in estuary marshes. Many of these strategies would alter more than one risk path, and all of them were identified as being helpful for other SFEP conservation goals.

Adaptation strategies	Green path	Purple path	Blue path	Other goals
Start restoration soon to achieve functions of mature marshes, including attainment of threshold elevations for organic accumulation, ahead of sea level rise	x			x
Plan for the temporal progression of habitats (e.g., by establishing habitats that will thrive under future climate conditions)	x			x
Plan for the spatial progression of restoration (e.g., consider impacts of broaching Suisun Marsh levees on downstream estuary restoration efforts)	x			x
Support resilience by restoring habitat complexity and facilitating high-energy parts of the system such as tides, wind-driven waves and freshwater flows		x	x	x
Practice integrated water management, including water conservation, as a priority	x	x		x
If it is not possible to make maintaining marsh salinity a top priority for Delta freshwater storage policies, plan for the restoration of tidal wetlands further up the estuary	x			x
Develop methods to move sediment into the bay, to keep pace vertically with sea level rise		x	x	x
Develop methods to reduce wave action on the front side of marshes			x	x
Adjust policies that prevent coarse sediment from entering the bay (e.g., for streams that don't support salmonids, change policies to allow an increase in sediment load)	x	x		x
Involve authorities in flood control districts to recouple streams to wetlands		x		x
Monitor change at the landscape scale to assess management effectiveness			x	x
Develop rapid response plans for catastrophes (e.g., levee breaks), with the political and scientific bases in place to respond properly			x	x

In its current program SFEP is already working on restoration, management and science activities that are related to these actions. The climate change vulnerability assessment and action plan can help to prioritize that work as well as point to how tailored projects can minimize risks.

EPA. 2012. *Vulnerability Assessments in Support of the Climate Ready Estuaries Program: A Novel Approach Using Expert Judgment, Volume I: Results for the San Francisco Estuary Partnership.* http://cfpub.epa.gov/ncea/global/recordisplay.cfm?deid=241556

TABLE 8A-1. POTENTIAL ADAPTATION ACTIONS FOR RISKS TO CLEAN WATER ACT GOALS

PART 1: CONTROL POINT AND NONPOINT SOURCES OF POLLUTION AND CLEAN UP POLLUTION

Adaptation actions	Warmer summers	Warmer winters	Warmer water	Increasing drought	Increasing storminess	Sea level rise	Ocean acidification
Create permitting rules that constrain locations for landfills, hazardous waste dumps, mine tailings and toxic chemical facilities			×		×	×	
Design new coastal drainage system					×	×	
Plug drainage canals for flood control					×	×	

PART 2: MAINTAIN AND IMPROVE ESTUARINE HABITAT

Adaptation actions	Warmer summers	Warmer winters	Warmer water	Increasing drought	Increasing storminess	Sea level rise	Ocean acidification
Composite systems—incorporate elements of two or more methods (e.g., breakwater, sand fill, planting vegetation)					×	×	
Create a regional sediment management (RSM) plan					×	×	
Create dunes along backshore of beach					×	×	
Create marsh by planting the appropriate species in the substrate						×	
Develop adaptive stormwater management practices				×	×		

PART 2: MAINTAIN AND IMPROVE ESTUARINE HABITAT (CONTINUED)

Adaptation actions	Warmer summers	Warmer winters	Warmer water	Increasing drought	Increasing storminess	Sea level rise	Ocean acidification
Establish rolling easements						x	
Fortify dikes					x	x	
Harden shoreline with breakwaters					x	x	
Harden shorelines with bulkheads					x	x	
Harden shorelines with revetments that armor the slope face of the shoreline					x	x	
Harden shorelines with seawalls					x	x	
Headland control					x	x	
Identify and protect ecologically significant areas such as nursery grounds, spawning grounds and areas of high species diversity	x	x	x				x
Incorporate consideration of climate change impacts into planning for new infrastructure				x	x	x	x
Incorporate wetlands protection into infrastructure planning					x	x	
Increase shoreline setbacks					x	x	
Install rock sills and other artificial breakwaters in front of tidal marshes along energetic estuarine shorelines					x	x	

PART 2: MAINTAIN AND IMPROVE ESTUARINE HABITAT (CONTINUED)

Adaptation actions	Warmer summers	Warmer winters	Warmer water	Increasing drought	Increasing storminess	Sea level rise	Ocean acidification
Incorporate coastal management into land use planning	x	x	x	x	x	x	
Integrated Coastal Zone Management (CZM)	x	x	x	x	x	x	
Land acquisition programs					x	x	
Land exchange programs					x	x	
Manage realignment and deliberately realign engineering structures					x	x	
Plant submerged aquatic vegetation to stabilize sediment and reduce erosion			x		x	x	
Preserve and restore structural complexity and biodiversity of vegetation in tidal marshes, seagrass meadows and mangroves					x		
Prohibit hard shore protection						x	
Redefine riverine flood hazard zones to tidal and riverine flow					x	x	
Remove hard protection or other barriers to tidal and riverine flow						x	
Remove shoreline hardening structures						x	

PART 2: MAINTAIN AND IMPROVE ESTUARINE HABITAT (CONTINUED)

Adaptation actions	Warmer summers	Warmer winters	Warmer water	Increasing drought	Increasing storminess	Sea level rise	Ocean acidification
Replace shoreline armoring with living shorelines					x	x	
Restrict or prohibit development in erosion zones					x	x	
Trap or add sand through beach nourishment					x	x	
Trap sand through construction of groins					x	x	
Use natural breakwaters of oysters to dissipate wave action and protect shorelines		x	x		x	x	
Wetland accretion by introducing sediment						x	
Wetlands migration						x	

PART 3: PROTECT AND PROPAGATE FISH, SHELLFISH AND WILDLIFE, INCLUDING CONTROL OF NONNATIVE SPECIES

Adaptation actions	Warmer summers	Warmer winters	Warmer water	Increasing drought	Increasing storminess	Sea level rise	Ocean acidification
Adapt protections of important biogeochemical zones and critical habitats as the locations of these areas change with climate	X	X	X	X	X	X	X
Connect landscapes with corridors to enable migrations	X	X				X	
Design estuaries with dynamic habitat boundaries and buffers	X	X			X	X	
Expand the planning horizons of land use planning to incorporate longer climate predictions				X	X	X	X
Purchase upland development rights or property rights				X	X	X	
Remove invasive species and restore native species	X	X	X	X			
Replicate habitat types in multiple areas to spread risks associated with climate change	X	X	X	X		X	X
Strengthen rules that prevent the introduction of invasive species	X	X	X				

PART 4: PROTECT PUBLIC WATER SUPPLIES AND RECREATIONAL ACTIVITIES, IN AND ON THE WATER

Adaptation actions	Warmer summers	Warmer winters	Warmer water	Increasing drought	Increasing storminess	Sea level rise	Ocean acidification
Create water markets	×	×		×	×	×	
Establish or broaden "use containment areas" to allocate and cap water withdrawal				×	×	×	
Incorporate sea level rise into planning for new infrastructure (e.g. sewage systems)						×	
Integrate climate change scenarios into water supply system	×	×	×	×	×	×	
Manage water demand	×	×	×	×	×	×	
Prevent or limit groundwater extraction from shallow aquifers	×	×	×	×		×	

Adapted from CRE's *Synthesis of Adaptation Options for Coastal Areas.*

Do you think it will work?

Do not worry about deeply assessing any action right now. This step is where you generate a list of actions for further investigation (in effect, hypotheses). At this scoping stage you can rely on your professional understanding of your system to ask yourself: is it reasonable to think the action would be effective at reducing the likelihood or consequence of the risk?

You will assess and affirm the risk reduction potential of the adaptation actions later in Step 8b. In this step you will simply confirm that by implementing an action (or combination of actions) you expect to change the risk's likelihood and/or consequence from high to medium or from high/medium to low. That is, you expect to move the risk at least one whole box in either dimension on your C/P matrix. If you are not going to be able to move a risk at least one box, do not spend a lot of time or money on actions that will not make much of a difference.

Every risk must have a mitigating action

If you cannot identify any potentially mitigating actions for a risk you should expand your conception of "plausible" to see if something rises up. Further, you might want to consider actions that are beyond your ability to implement alone. If this is a high-consequence risk to your organization, then you might need to take more extreme steps to mitigate it.

Ultimately, however, if you cannot find an action (or combination of actions) that will move a risk in your C/P matrix from high to medium or from high/medium to low, in at least one dimension, then you really need to review your decision of trying to mitigate it. This is another case in which more information has become available and you should return to an earlier step—Step 7 in this case—so you can decide on another approach for this risk. Note that if you cannot think of any way to mitigate the risk, then transferring the risk to another party to mitigate probably is not a viable option either. You probably need to accept or avoid this risk.

To Get Started

Take a look at each of the risks you identified for mitigation and determine if any diagrams you used to help with risk identification in Step 3 or if the generic conceptual models in Appendix D will be helpful for understanding the risk pathways.

TABLE 8A-2. SELECTION OF ADAPTATION ACTIONS

Risk selected for mitigation	Potential adaptation action (one or more for each risk)	Could the action reduce likelihood (by itself or in combination with another action)? Yes/No	Could the action reduce consequence (by itself or in combination with another action)? Yes/No
1.			
2.			
3.			
n.			

Additional Resources

Also see Appendix B.

EPA resources to identify additional adaptation actions for consideration

Climate Ready Estuaries. 2009. *Synthesis of Adaptation Options for Coastal Areas.*
http://www.epa.gov/cre/

Climate Ready Water Utilities. 2012. *Adaptation Strategies Guide for Water Utilities.*
http://water.epa.gov/infrastructure/watersecurity/climate/upload/epa817k11003.pdf

EPA and NOAA. 2011. *Achieving Hazard-Resilient Coastal and Waterfront Smart Growth: Coastal and Waterfront Smart Growth and Hazard Mitigation Roundtable Report.*
http://coastalsmartgrowth.noaa.gov/pdf/hazard_resilience.pdf

Using conceptual models (also see Appendix D)

EPA. 2012. *Vulnerability Assessments in Support of the Climate Ready Estuaries Program: A Novel Approach Using Expert Judgment, Volume 1: Results for the San Francisco Estuaries Partnership.*
http://ofmpub.epa.gov/eims/eimscomm.getfile?p_download_id=505773

EPA. 2012. *Vulnerability Assessments in Support of the Climate Ready Estuaries Program: A Novel Approach Using Expert Judgment, Volume 2: Results for the Massachusetts Bay Program.*
http://ofmpub.epa.gov/eims/eimscomm.getfile?p_download_id=505771

EPA. 2012. CADDIS: The Causal Analysis/Diagnosis Decision Information System.
http://www.epa.gov/caddis/index.html

National Park Service. Integrated resource management applications.
http://science.nature.nps.gov/im/index.cfm
https://irma.nps.gov/App/

Additional adaptation options

NOAA. 2010. *Adapting to Climate Change: A Planning Guide for State Coastal Managers.* Step 3.2, pp. 46–96.
http://coastalmanagement.noaa.gov/climate/docs/adaptationguide.pdf

Center for Climate Strategies. 2011. *Center for Climate Strategies Adaptation Guidebook: Comprehensive Climate Action.* Appendix 3.
http://www.climatestrategies.us/library/library/view/908

American Rivers and Natural Resources Defense Council. 2012. *Getting Climate Smart: A Water Preparedness Guide for State Action.* pp. 65–99.
http://www.nrdc.org/water/climate-smart/files/getting-climate-smart.pdf

National Wildlife Federation. 2014. *Green Works for Climate Resilience: A Guide to Community Planning for Climate Change.*
http://www.nwf.org/~/media/PDFs/Global-Warming/Climate-Smart-Conservation/2014/green-works-final-for-web.pdf

California Emergency Management Agency and California Natural Resources Agency. 2012. *California Adaptation Planning Guide: Identifying Adaptation Strategies.*
http://www.ca-ilg.org/sites/main/files/file-attachments/apg_identifying_adaptation_strategies.pdf

EPA Office of Water, Oceans and Coastal Protection Division

What Is "Selecting Adaptation Actions"?

In this step, the adaptation actions you found in Step 8a will be screened to determine whether they are desirable and still candidates for implementation. Actions that pass your screening tests will be ranked according to risk-reduction potential and the top actions will be selected for implementation.

Objective of This Step

The objective of this step is to screen candidate adaptation actions and select the ones your organization will move forward for implementation.

Process

In Step 8a, you identified a preliminary list of adaptation actions that you believe would disrupt a risk and lead to a reduction of its likelihood or consequence. Some mitigating actions will be obvious winners, but others may have some flaws. In this step, you will further assess each adaptation action and use Table 8b-1 to record your results.

When you have a set of good actions that pass your screening tests, then you will once again turn to your vulnerability assessment. Your C/P matrix will help you to rank your actions based on their risk reduction potential. Then you will select the actions that will achieve the most risk reduction for your organization based on your available resources.

Criteria to assess actions

In addition to risk reduction potential, which you conditionally affirmed in the last step, the WORKBOOK uses five other sets of questions (*America's Climate Choices Summary Report*, p. 46) to apply to the pool of potential actions.

- Feasibility and effectiveness.
- Cost and cost-effectiveness.
- Ancillary costs and benefits.
- Equity and fairness.
- Robustness.

Risk reduction potential

You already considered the risk reduction potential of actions in Step 8a by affirming that you expect each adaptation action (by itself or by a combination of actions) to move a risk at least one box on your consequence/probability matrix. Do you still believe that?

Feasibility and effectiveness

- Is the action a proven strategy? Are there other places that have successfully implemented this action?
- Do you have enough time to implement it to prevent risks from occurring?
- Is it politically feasible?
- Do you have, or can you get, authority or permission to implement it?
- Is it something your community/stakeholders would accept?

Cost and cost-effectiveness

- Is it affordable? Which category does the cost of the adaptation action fit?
  - Minor.

Managing for change

"Beyond 'managing for resilience,' the Nation's capability to adapt will ultimately depend on our ability to be flexible in setting priorities and 'managing for change.' Prioritizing actions and balancing competing management objectives at all scales of decision making is essential, especially in the midst of shifting budgets and rapidly changing ecosystems....Over time, our ability to 'manage for resilience' of current systems in the face of climate change will be limited as temperature thresholds are exceeded, climate impacts become severe and irreversible, and socioeconomic costs of maintaining existing ecosystem structures, functions, and services become excessive. At this point, it will be necessary to 'manage for change,' with a re-examination of priorities and a shift to adaptation options that incorporate information on projected ecosystem changes."

—Executive Summary, p. 4

USGCRP. 2008. *Preliminary Review of Adaptation Options for Climate-Sensitive Ecosystems and Resources.* Synthesis and Assessment Product 4.4.

— Similar to a municipal public works project (i.e., can afford this in the budget or by bonding or fundraising, or by extending it over multiple budget years).

— Very expensive, requiring external assistance.

— Overwhelming and impossible.

• Is the cost reasonable for the risk reduction an action would produce?

• Did you consider if there are long-term maintenance costs?

• Does the action prevent other future costs or damages?

• Will the action generate positive economic benefits?

Some of the most effective options may also be the most expensive. The converse—the cheapest actions might be the least effective—also holds. Later in this step you will consider how your available resources affect how many actions you can implement and how much risk reduction you can achieve.

Ancillary costs and benefits

• Is the action maladaptive?

— Does action in one sector/place cause problems in another sector/place?

— Will action to reduce some risks encourage people to take other risks they would have avoided otherwise?

— Does the action close off other options or lock the future onto one path?

• Co-benefits—does an action create positive side effects that help with other non-climate problems or with more than one climate change risk?

• Sustainability—is there a balance between the social, environmental and economic costs and benefits of this action?

— Are needs of present and of future generations considered?

— Are there adverse impacts on the environment, ecosystem functions or ecosystem services?

Equity and fairness

• Does it align with your ethics and principles?

• Does it cause a minority population or low-income population to bear disproportionately high and adverse effects?

Sustainability

"'[S]ustainability' and 'sustainable' mean to create and maintain conditions, under which humans and nature can exist in productive harmony, that permit fulfilling the social, economic, and other requirements of present and future generations."

—Executive Order 13514 of October 5, 2009

The ultimate goal of adaptation is to create a sustainable place (otherwise adaptation actions are just patches and short-term fixes). You want to create places that will be resilient to the risks that future climate changes will pose.

The National Research Council recommends the "Three Pillars" approach of "Social," "Environment," and "Economic" dimensions of sustainability. The intent is to create an integrated, forward-thinking approach that takes all of those elements into account.

National Research Council. 2011. *Sustainability and the U.S. EPA.*

A "disproportionately high and adverse" effect or impact...

"...(1) is predominately borne by any segment of the population, including, for example, a minority population and/or a low-income population; or (2) will be suffered by a minority population and/or low-income population and is appreciably more severe or greater in magnitude than the adverse effect or impact that will be suffered by a non-minority population and/or non-low-income population."

EPA. 2004. *Toolkit for Assessing Potential Allegations of Environmental Injustice.* EPA 300-R-04-002.

- Are many asked to pay for benefits that accrue to just a few?
- Does the action limit the adaptation options of other places or sectors?

Robustness

- Given inherent uncertainty about the future, will the action perform under a range of potential conditions?
- Is it a flexible adaptation action? Can it be modified at a future date if the climate changes differently than expected?
- Are you putting all your eggs in one basket, or is this action part of a coordinated set of actions?
- Is it a no-regrets action? Is it just a good idea regardless of how climate might change?
- Is it a win-win action? Would it also be a good strategy for achieving other desired social, environmental or economic results?
- Are there any circumstances where you would become sorry you implemented this action?

These six tests can have many parts and the tests are not simple pass/fail questions. You are trying to find actions that are desirable as well as effective—and you want to avoid problematic actions. These screening tests are really about deciding whether an action is unsuitable or appropriate for your organization. Record your conclusions in Table 8b-1 and note whether you are willing to proceed with each action.

TABLE 8B-1. EVALUATION OF ADAPTATION ACTIONS

Adaptation actions	Risk reduction potential	Feasibility and effectiveness	Cost and cost-effectiveness	Ancillary costs and benefits	Equity and fairness	Robustness	Appropriate to proceed with this action? (yes/no)
1.							
2.							
3.							
4.							
5.							
6.							
7.							
8.							
n.							

Reviewing the screening assessment

Not every action will pass all the tests. You truly do not want to implement actions that will not work, that you cannot afford, or that cause other problems, but be alert for ways to compensate for those problems.

Which tests an action fails matter.

- If you disqualified an action because you cannot afford it, you might be able to transfer that risk to others who can pay the costs. Alternatively you may need to focus attention on finding the funds.

- If the impediments are political or structural, then your first acts toward reducing a risk may be to find ways to overcome those obstacles.

- If an action generates unwanted effects, then those tradeoffs might be mitigated too.

You may not be able to find any actions for some risks.

- If you cannot find a single mitigating action for a risk that passes your screening tests and that you want to implement, then a potential partner probably will not find one either, so transferring this risk may not work. You probably need to accept or avoid the risk.

- Keep track of actions that failed one or more of the screening tests because conditions may change over time and these actions may become more attractive in the future.

The best actions will pass each screening question. Actions that do satisfactorily on your tests will move forward for further consideration of their risk reduction potential in the rest of this step.

Risk-based ranking of actions

This WORKBOOK assumes (as expressed in the Introduction) that users will have to prioritize response actions because not all can be implemented. The mitigating actions that pass the six criteria will address a variety of risks that will affect different goals, that will occur in different habitat types, and that range from high-impact risks (red in your matrix) to low-impact risks (green).

In a risk management process, the best actions will be the ones that achieve the most risk reduction. Your task now is to sort all of your screened actions according to how much of your climate change risk they will mitigate.

For each action, you first need to return to the C/P matrix and see whether the primary target of the action is a red, yellow or green risk. Do a rough sort of your list so that actions that respond to red risks are at the top, and actions that respond only to green risks are at the bottom (Figure 8b-2, left).

What follows is a set of decision-making guidelines that will help you refine the rough sorting. Some actions are more attractive because they respond to your organization's total risk better than others, and they should float higher on your list. Note that a group of actions can be assessed as one coordinated set. Your C/P matrix and the information you developed about the action in Step 8a will help you to construct the ranking.

Actions that address red risks

- Responding to red risks should be at the top of the list. You said these risks are highly likely to happen and will have high impacts when they do. Actions that would mitigate a red risk should get priority over responses to a yellow risk.

- Actions that reduce more than one risk are more attractive than comparable ones that reduce only one risk. Be sure the main target risk is making real movement to the lower left of your matrix. Do not do a little bit for a red risk just because it also does a little for a yellow risk: focus on meaningfully reducing risk.

- Actions that would move red risks closer to the lower left corner of your matrix are more attractive than comparable actions that would leave them further away. But do not spend a fortune to mitigate risks down to zero if getting them into the green zone is sufficient.

- Do not duplicate efforts unless you intend to. You should not expend resources to implement more than one solitary action to achieve the same risk reduction, but if an action is part of a coordinated set of risk reducing actions, that is fine. If an action is part of a planned set of redundant actions, that is fine too.

Actions that address yellow risks

- Yellow risks should not be the primary target of your efforts unless you have no red risks. Look for opportunities to move yellow risks into the green zone as part of your work on other red risks.

- You may have an opportunity to take an action that would reduce a few yellow risks instead of taking another action that would reduce one red risk. This is a judgment call based on the magnitude of all the contextual benefits each choice could achieve. However, do not work to clean out the yellow zone while leaving red risks intact.

Actions that only address green risks

- Actions that respond only to green risks should be at the bottom of your list. You have decided these risks are less likely to occur and would not matter as much if they did. You probably should not dedicate any resources to them that would take away from responding to yellow or red risks.

- Actions that have negligible costs (in time and money) and can be taken as a part of other normal business are more attractive than other actions aimed at green risks.

- Instead of launching a project aimed at a green risk, look for opportunities to fine-tune an action you will use to address a red or yellow risk. Maybe those actions can incorporate a small component that further reduces green risks.

After you assess your pool of actions, you should have a relative ranking (like the example on the right side of Figure 8b-1) of the whole set of actions based on how much of your organization's overall risk they reduce. This is a risk-based ordering of actions.

Selecting actions for implementation

You are setting out to reduce the overall climate change risk to your organization. A risk-based ordering of actions means that each action (or coordinated set of actions) reduces more risk than the item below it on the list. If you moved actions to be out of sequence on the list (either further up or further down), then some actions that reduce less risk would appear higher on your list than actions that reduce more risk.

If you need to prioritize response actions because you will not be able to implement them all, the next thing is to divide your actions into two tiers. Tier 1 actions are the ones your organization has the capacity to start on within your workplan cycle. Tier 2 is the waiting list. Taking into account your organization's financial and human resources, you should draw a line separating the actions into Tier 1 and Tier 2 as far down from the top of the list as you can in order to maximize the number of actions in Tier 1 (Figure 8b-2).

If Tier 1 reaches down to an action that you cannot afford but you can afford more actions below that particularly expensive one, then: (1) note that one action is too costly and that you are skipping it in this

> ### Priorities
>
> You are free to implement any actions you like: this ordered list is not a mandatory plan. The list is arranged by how much risk actions reduce. Remember that your organization said the highest risks had the highest potential to keep you from your goals. Be aware that the implication of choosing to work on smaller risks or pursuing actions with less mitigation potential is that the consequences of higher risks may appear.

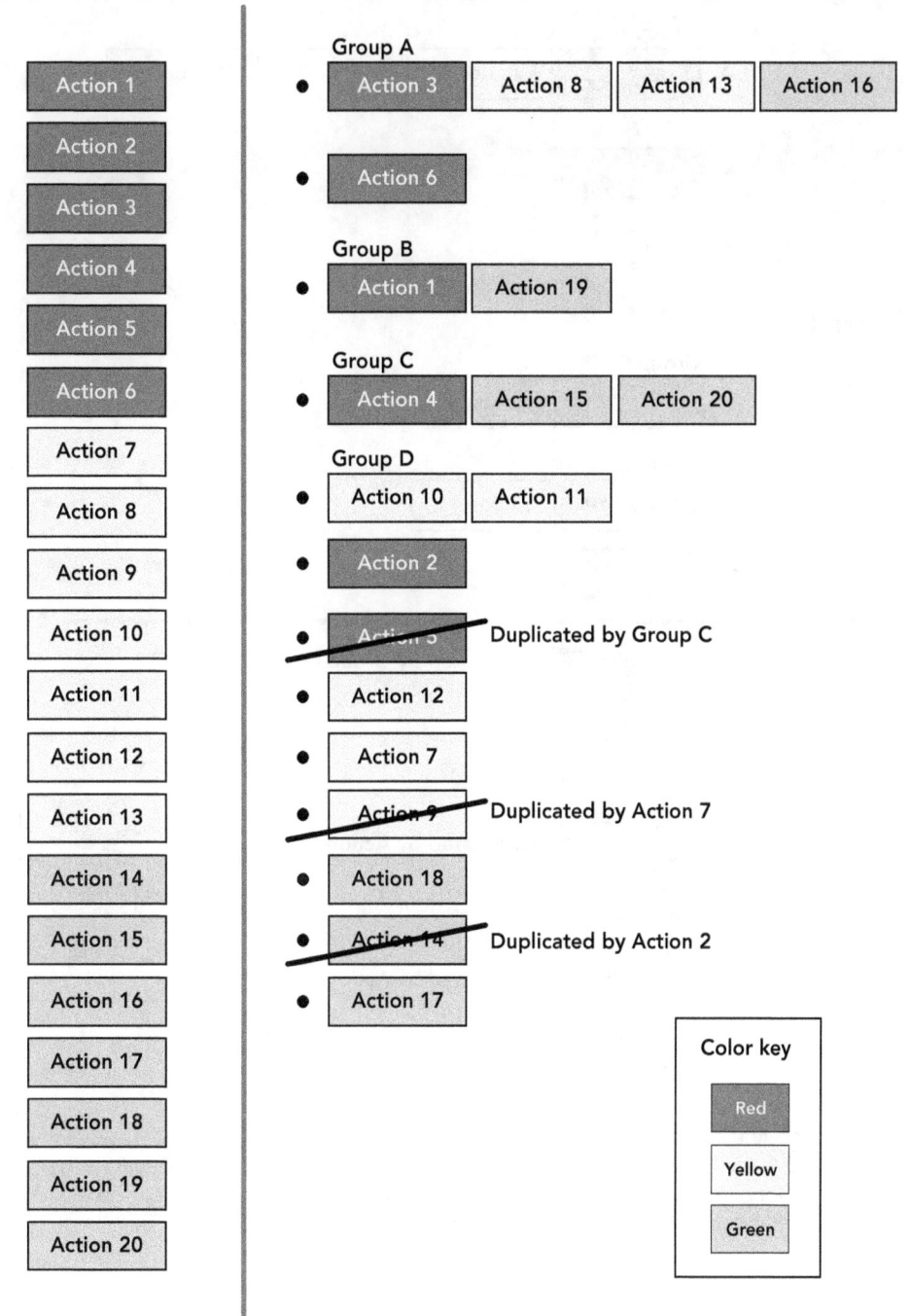

FIGURE 8B-1. **Left:** A rough initial sort places actions aimed at red risks at the top (in no particular order) and those aimed principally at green risks at the bottom. **Right:** Further refine the list so that from top to bottom, actions (or groups of actions) go from most risk reduction to least. If any actions duplicate the risk reduction of other actions, strike the duplicative ones and make a note (while keeping their place in the ordered list). An ordered list like the one on the right is your risk-based ranking of actions, in which every action (or group of actions) reduces overall organizational risk more than the actions below it.

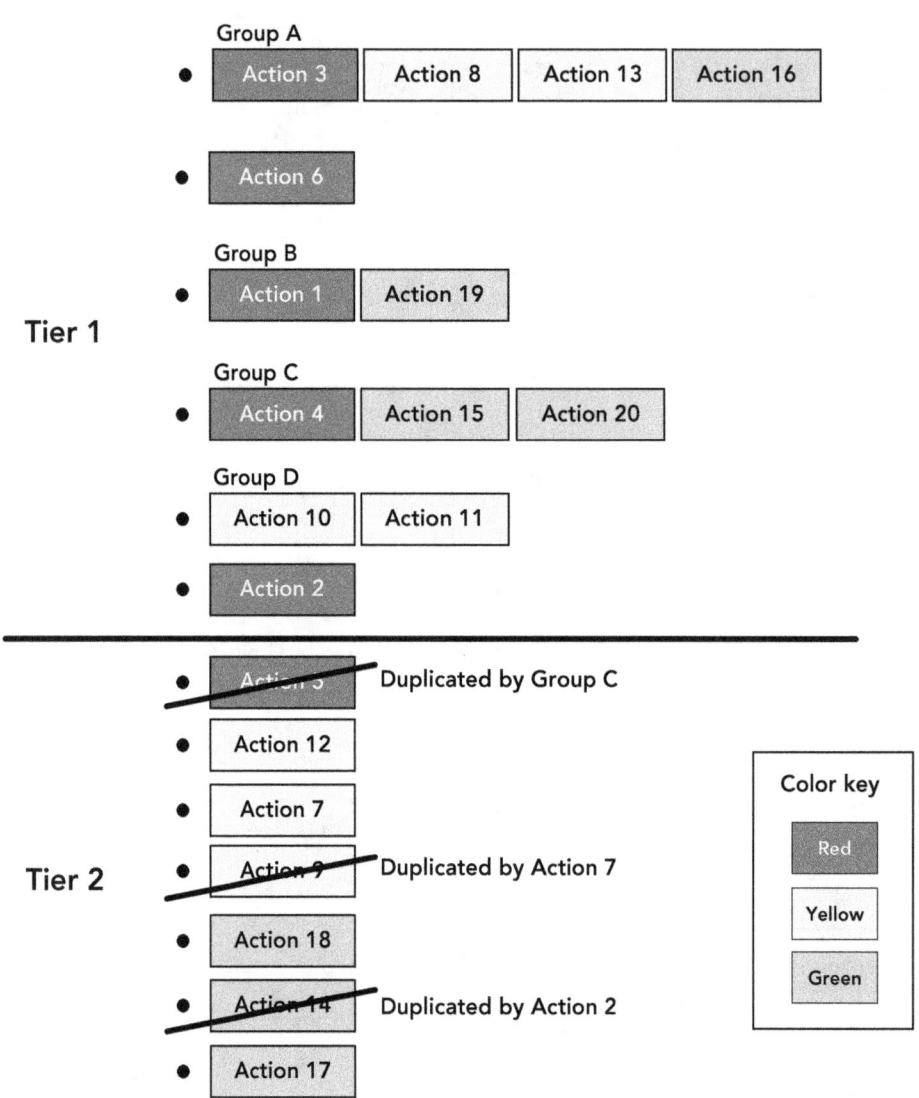

FIGURE 8B-2. Place a line as far down from the top of a list like Figure 8b-1 as your resources will allow. Tier 1 actions (above the line) will move forward for implementation, while Tier 2 actions (below the line) will not move forward right now. This is the plan that will get your organization the most risk reduction you can achieve using available resources.

work planning cycle—but don't change its place in the ordered list—and (2) move the line down so that as many actions as you can afford are in Tier 1.

The actions in Tier 1 generate the most risk reduction you can realize at this time. Tier 1 actions will move into Step 9 for implementation. The actions that are beyond your current capacity will be in Tier 2 and will not move forward right now. Actions that you skip are technically in Tier 2, but keep their place in the ordered list.

Review your risks

Any risk that you selected for mitigation in Step 7 that does not have an associated Tier 1 action at the end of this step is not going to be mitigated at this time. You need to return to Step 7 and change the approach you selected for it. If you have an action for the risk in Tier 2 and your organization still intends to mitigate the risk yourself some day, then note that.

Risks that are being accepted

In Step 7—Risk Evaluation: Deciding on a Course, you may have decided to accept some risks. You may have elected to continue with business as usual and run a risk, dealing with the impact if/when it does occur. Your vulnerability assessment may have concluded that some risks are unlikely, or that their consequences are small. You may have decided that some risks are far enough away in time that they are not immediate problems and you can address them effectively later. Indeed Figure 7-3 suggests that when accepting a risk is a reasonable choice (e.g., for reasons like these) then you should do that.

However, you should take a closer look at the risks that you once thought should be mitigated, but now you will not be able to. These may have moved from the mitigation approach to the accept approach (Figure 8b-3) because:

- you could not find a potential mitigating option (Step 8a);
- you could not find a mitigating action that passed your screening tests (Step 8b); or
- you do not have the resources to implement a corresponding mitigation action (Step 8b).

The risks that are now in the accept group, but really need to be mitigated, are existential problems. When you determine that a number of risks are highly likely to occur and will have high consequences when they do, and you are currently unable to do anything about it, it is time for critical thought about your mission. If business as usual is unsustainable and you cannot find a way to maintain it, then your organization may need to have serious discussions about "managing for resiliency" vs. "managing for change."

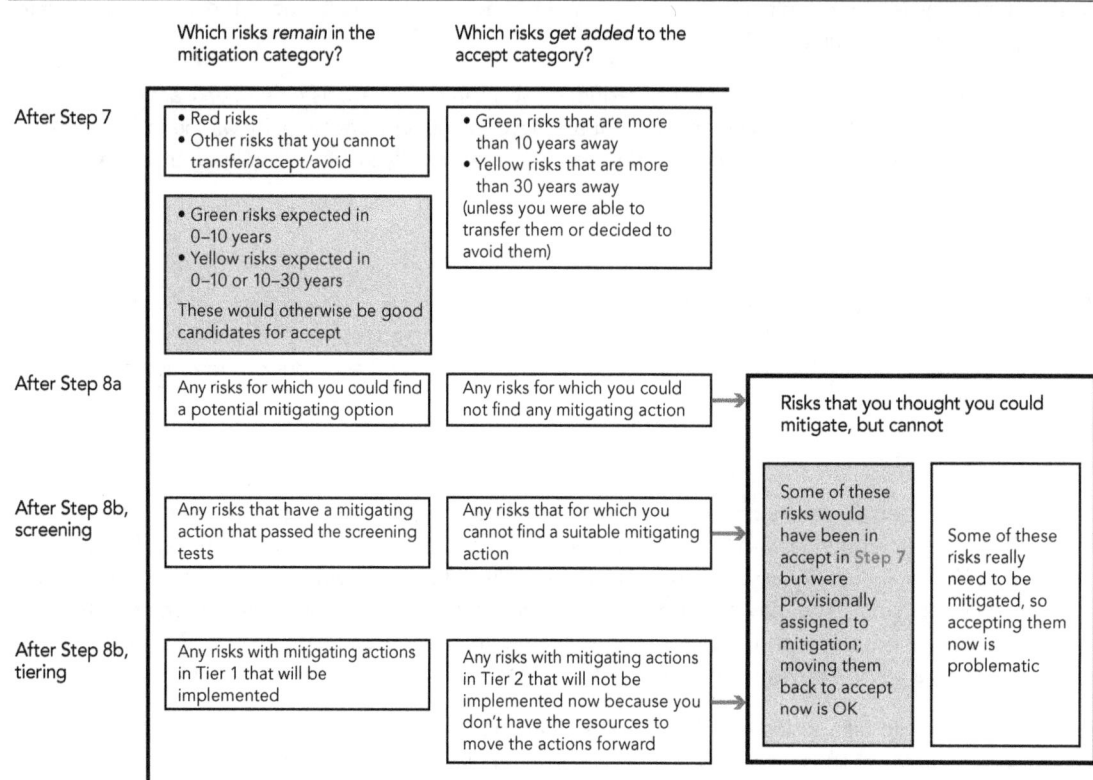

FIGURE 8B-3. In Step 7, Step 8a and here in Step 8b, a series of decisions may have affected whether risks would be assigned to either the mitigate approach or the accept approach. Some of the risks that have ended up in the accept approach are there because for some reason you will not be able to mitigate them—even if they are real threats and need to be mitigated.

To Get Started

Start your screening with the actions in Table 8a-2 that will reduce both the likelihood and consequence of your risks.

Additional Resources

Also see Appendix B.

Resources for assessing adaptation actions

National Research Council. 2011. *America's Climate Choices: Summary Report.*
http://www.nap.edu/catalog.php?record_id=12781

National Research Council. 2010. *America's Climate Choices: Adapting to the Impacts of Climate Change.*
pp. 137–144.
http://nas-sites.org/americasclimatechoices/sample-page/panel-reports/panel-on-adapting-to-the-impacts-of-climate-change/

National Wildlife Federation. 2011. *Restoring the Great Lakes' Coastal Future.* pp. 37–38.
http://www.nwf.org/~/media/PDFs/Global-Warming/Climate-Smart-Conservation/NWF_Restoring_the_Great_Lakes_Coastal_Future_090211.ashx

New York City Panel on Climate Change. 2010. *Adaptation Assessment Guidebook.* Appendix B.
http://onlinelibrary.wiley.com/doi/10.1111/j.1749-6632.2010.05324.x/pdf

ICLEI. 2007. *Preparing for Climate Change: A Guidebook for Local, Regional, and State Governments.*
Chapter 10.3, p. 97.
http://www.icleiusa.org/action-center/planning/adaptation-guidebook

National Wildlife Federation. 2011. *Scanning the Conservation Horizon: A Guide to Climate Change Vulnerability Assessment.* Chapter 6, pp. 79–80.
http://www.nwf.org/~/media/PDFs/Global-Warming/Climate-Smart-Conservation/NWFScanningtheConservationHorizonFINAL92311.ashx

John McShane, EPA Office of Water

What Is "Preparing and Implementing an Action Plan"?

Preparing and implementing an action plan means that you will designate responsible parties for each Tier 1 adaptation action identified in Step 8b. You will then charge the responsible parties with developing and implementing project plans.

Objective of This Step

The objective is to create a plan to ensure that each action identified as Tier 1 is moving forward and that risks selected for mitigation are being reduced over time.

Process

You are ready to begin writing your action plan. You know your risks, you know which actions will help reduce your risks, and you know how you want to proceed. In this step, you will designate responsible parties and record information to help you track the implementation of each Tier 1 adaptation action you identified in Step 8b.

Concept approval

In Step 7, you may have worked with your partners to negotiate a transfer of certain risks. You may also have decided to run some risks by choosing to accept them. You may have decided to avoid some risks. You decided to try to mitigate the remaining risks. You also confirmed with your key decision-makers that they concur with the high-level approach for each risk. In Step 8b, you selected promising adaptation actions for all the risks you intend to mitigate, and you are proposing to move that set forward for project design and implementation. Before you go further with your action plan, this is another opportunity to run it by your organization board of directors, management committee, staff, and the people who are regularly involved in decisions for their concurrence about how you propose to manage your risks.

A planning-level plan

In this last planning-level step, you will create the mechanisms to ensure that you make progress. You want to create two risk management tracking systems. One tracks the actions; the other tracks the risks. You will create two tables that start as inverses of each other. Each table should be updated regularly.

The first table (Table 9-1) tracks the actions and notes which risks they are associated with as well as who has responsibility for implementing them. An action may address more than one risk.

The second table (Table 9-2) keeps track of your risks and notes what actions are being used and when they are completed. A risk may be affected by more than one action. You should also track any activities associated with risks transferred to another organization.

From planning level to project level

The WORKBOOK is designed to help you create a planning document for managing risks at a watershed scale. Once a responsible person is named for implementing an action, what comes next is beyond the scope of this WORKBOOK.

Assigning responsibility to project managers who will act to implement the actions that lower climate change risks might be this step's most important outcome right now. When people are named as responsible parties, they

Opportunities

If you identified any opportunities in Step 3: Risk Identification, this is the time to pick them back up. You want to make sure to find strategies that take advantage of those positive outcomes and maximize their benefit.

Approaches for opportunities

Exploit—Positive impacts to your goals are possible, provided the necessary resources to realize the benefits exist.

Share—Outsourcing and making better use of external partnerships may be required in order to capture (all of) the opportunity.

Enhance—Affecting key drivers to increase the expected value of the opportunity.

Accept—This approach indicates that the program team has decided not to change program plans and will accept the opportunities or benefits as they occur.

Adapted from: Project Management Institute. 2008. *The Standard for Program Management.* Chapter 11.4.2, Plan program risk responses: Tools and techniques.

assume all the duties of a project manager, among them defining the project scope, completing on time, and staying within resources.

Project managers may establish project teams; do detailed site reconnaissance; hire experts; evaluate pros and cons of specific techniques; create construction documents; get a staffing or financing plan, or a cooperative agreement, in place; file for permits; hire a construction contractor; or initiate other related activities. It may be helpful at this point to turn to standard project management tools as well as to the resources in Appendix C, which are designed for those who are doing climate change work at finer resolutions.

Monitoring

To understand whether a project has successfully met its objectives, you may need to include post-project monitoring activities in the scope of work.

To Get Started

In Table 9-1, list each Tier 1 adaptation action (from Step 8b) and identify which risk(s) it addresses; then do the inverse for Table 9-2.

TABLE **9-1.** TRACKING SELECTED ACTIONS

Adaptation action	Risk(s) addressed	Responsible party(ies)	Next steps	Reporting frequency
1.				
2.				
3.				
n.				

TABLE **9-2.** TRACKING RISK REDUCTION

Risk selected for mitigation	Action(s) employed/ completed
1.	
2.	
3.	
n.	

Additional Resources

Also see Appendix B.

Using vulnerability assessment results

National Wildlife Federation. 2011. *Scanning the Conservation Horizon: A Guide to Climate Change Vulnerability Assessment.* Chapter 6, pp. 79–80.
http://www.nwf.org/~/media/PDFs/Global-Warming/Climate-Smart-Conservation/NWFScanningtheConservationHorizonFINAL92311.ashx

EPA Region 9 and California Department of Water Resources. 2011. *Climate Change Handbook for Regional Water Planning.* pp. 7-1–7-17.
http://www.water.ca.gov/climatechange/CCHandbook.cfm

EPA. 2005. *Community-Based Watershed Management: Lessons from the National Estuary Program.* Chapter 5.
http://water.epa.gov/type/oceb/nep/handbook.cfm

Partnership for the Delaware Estuary

What Is "Monitoring and Review"?

This is the process of checking in on the effectiveness of the mitigating actions as they are implemented. Check back on the risks that were identified for the transfer, accept, or avoid approaches too, in case additional information has become available and a different approach is more appropriate. You will also stay current with climate change issues so that your vulnerability assessment does not get out of date.

Objective of This Step

The objective of this step is to monitor and review your vulnerability assessment and action plan and to return to earlier steps to update them if necessary.

Process

In Step 9 the planning-level work was done and actions were turned over to project managers for implementation. In this step you return to the planning level to monitor and review progress and your updated situation.

As risk mitigation projects progress or are completed you will have new consequences and likelihoods for your risks. As new scientific information emerges about the magnitude of stressors or about what the impacts of climate change will be, you may discover new risks or the consequences and likelihoods of your current risks may change. Your organization's context or partners may also change over time in ways that cause you to re-evaluate how climate change will affect what you are trying to achieve.

Risk management methodologies (including ISO 31000) typically call for recycling back through the process once the end of an iteration is reached. Frequent monitoring and review will keep your plan fresh and prevent you from having to start from the beginning later on. It may make sense for your monitoring and review to be conducted at the same time as regularly scheduled strategic planning or management review of the organization.

Monitoring and review

Monitoring and review focuses on the implementation of your action plan. Conditions are always changing, so incorporate any new information to ensure the best plan is still in place. Considerations for monitoring and review include:

Context: The context of your organization has been used to help you identify risks relating to your goals, as well as determine the unique circumstances under which you operate and will be able to address your risks.

As time passes, your context may change, so your plan will need to be adjusted to suit these changes. Your organization may take on new challenges and have new goals. Your budget or personnel may change. Perhaps the political situation in which you operate evolves. Or maybe you identify more partners to transfer risks to—freeing up more resources and time for you to address other risks.

In any case, you will need to review and revise your action plan to reflect changes in your regulatory, political and financial context and ensure the plan reflects your current context.

> ### Questions to consider
>
> How often should the planning team meet to review the plan and discuss its progress?
>
> How will the planning team keep key decision-makers apprised of progress and obstacles?
>
> How will progress be communicated to other stakeholders outside the planning team?
>
> How should progress be reported?
>
> Were there any unintended consequences due to implementation (good or bad)?
>
> Are there any new stressors or challenges?
>
> Adapted from: NOAA. 2010. *Adapting to Climate Change: A Planning Guide for State Coastal Managers.*

Climate change: New climate change science will likely emerge after you have completed your adaptation plan. Information about magnitude of projected changes and new knowledge about expected impacts will continue to be improved. Incorporate new information (e.g., science, data) as it becomes available so that your plan stays current and effectively addresses your risks. The U.S. National Climate Assessment is updated from time to time: updated releases of the assessment might be an opportune time to revisit the assumptions you made about stressors and risks.

Risks: Your risks will likely evolve over time as well, especially as climate changes accumulate or interact. What used to be a moderate risk could become a more severe risk over time, one you feel needs to be addressed sooner rather than later. Your own mitigating actions hopefully change how you would assess a risk, too. Update your plan to reflect changes and continue through the subsequent steps to revise your action plan.

Action plan: Your action plan is meant to be a living document. You should update it often, especially if anything changes regarding your context, if new climate change science emerges, if new risks are identified, or if current risks become a more imminent danger than when you first addressed them. It is also useful to review the action plan to ensure that actions being implemented are having the intended effect (and not having unintended negative effects). If you find your identified actions are not the best way to address a particular risk, then cycle back through the action planning process for that risk.

Communications: Results of monitoring and review should be recorded and reported internally and externally as appropriate. Keep stakeholders, partners and funders informed about your progress and any changes to your plans. Your completed action plan should also be used to communicate to others what your organization is doing to address your climate change risks.

> ## Adaptive management
>
> "Adaptive management is an iterative approach that seeks to improve natural resource management by testing management hypotheses and learning from the results. A management action can have the desired effect on the distribution and abundance of the target species. However, depending on the type of management action, there can also be a number of unintended consequences. Adaptive management provides a research/management tool to assess the frequency and intensity of unintended effects. It is an approach that is useful in situations where uncertainty about ecological responses is high, such as climate change."
>
> USGCRP. 2008. *Preliminary Review of Adaptation Options for Climate-Sensitive Ecosystems and Resources.* Synthesis and Assessment Product 4.4. Chapter 5.4.1.

To Get Started

Working within your organization and with key stakeholders and partners, determine a monitoring schedule for each of your mitigation projects (which could be the same as your project review period for other non-climate work), as well as when a review of the vulnerability assessment will occur (which could simply be the passage of a set period of time).

Additional Resources

Also see Appendix B.

EPA resources for monitoring project results

EPA. 2008. *Handbook for Developing Watershed Plans to Restore and Protect Our Waters.* pp. 12-10–12-12.
http://water.epa.gov/polwaste/nps/handbook_index.cfm

EPA. 2010. *Measurement Tips & Resources for Community Projects.*
http://www.epa.gov/care/library/CARE_Measurement_Tips.pdf

Resources for tracking progress

Association of Fish & Wildlife Agencies. 2009. *Voluntary Guidance for States to Incorporate Climate Change into State Wildlife Action Plans & Other Management Plans.* pp. 20–21.
http://www.fishwildlife.org/files/AFWA-Voluntary-Guidance-Incorporating-Climate-Change_SWAP.pdf

NOAA. 2010. *Adapting to Climate Change: A Planning Guide for State Coastal Managers.* Chapter 6, Step 4.4, pp. 104–105.
http://coastalmanagement.noaa.gov/climate/docs/adaptationguide.pdf

Appendices

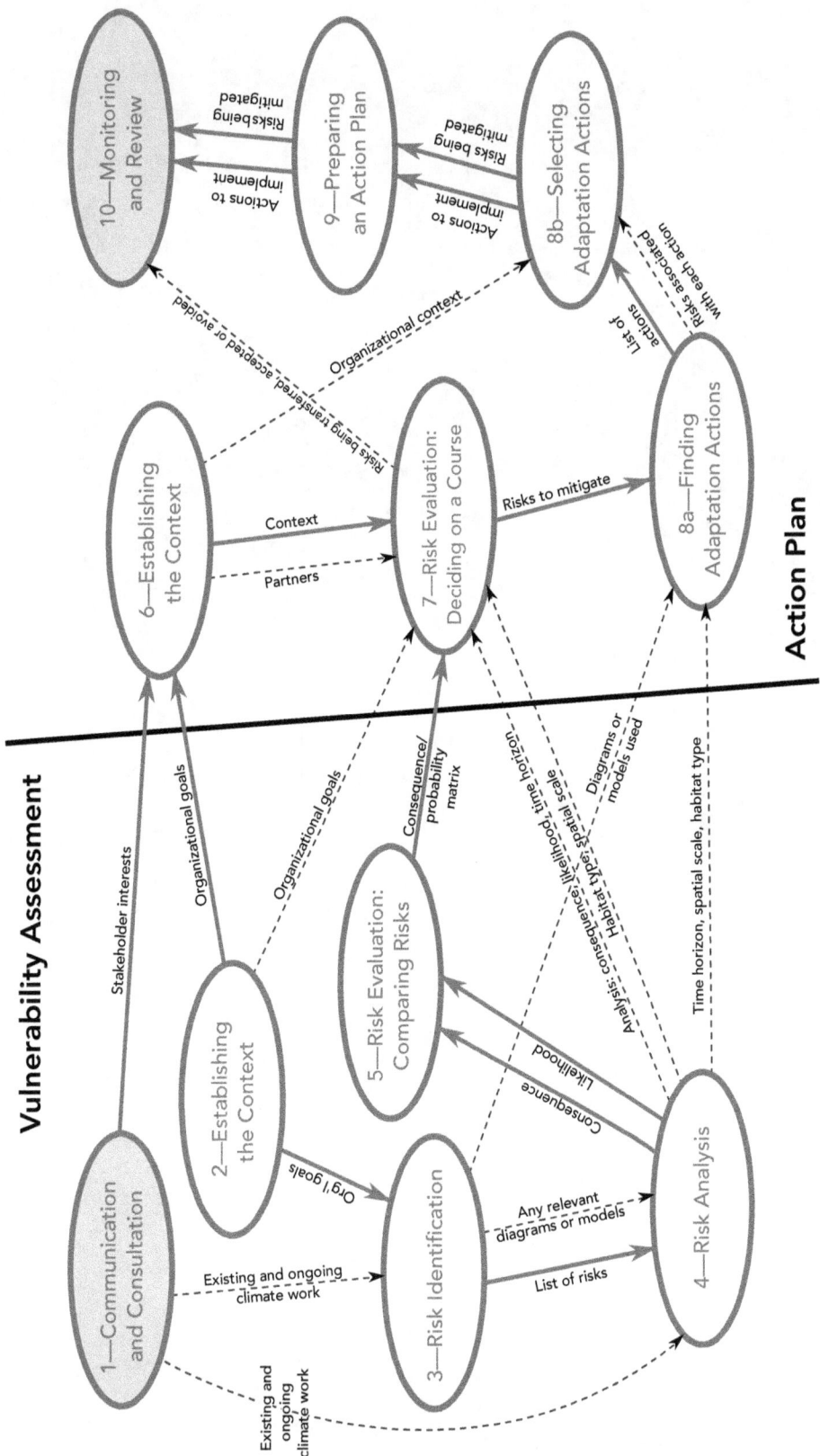

FIGURE A-1. A detailed map of information development and use in the WORKBOOK. Thicker blue arrows indicate the primary information flow from Step 1 through Step 5 of the vulnerability assessment, and from Step 6 through Step 10 of action planning. Secondary flows of information are shown with dashed arrows. Results from earlier steps are continually being used. Communication and consultation happen throughout the process. Monitoring and review keep all the steps up to date.

These cross-references are provided for those who have already used these other resources and for people who would like additional procedural help.

- NOAA. n.d. *Roadmap for Adapting to Coastal Risk.*
  https://www.csc.noaa.gov/digitalcoast/training/roadmap

- ICLEI. 2007. *Preparing for Climate Change: A Guidebook for Local, Regional, and State Governments.*
  http://www.icleiusa.org/action-center/planning/adaptation-guidebook

- NOAA. 2010. *Adapting to Climate Change: A Planning Guide for State Coastal Managers.*
  http://coastalmanagement.noaa.gov/climate/docs/adaptationguide.pdf

This Workbook	NOAA: *Roadmap for Adapting to Coastal Risk*	ICLEI: *Preparing for Climate Change*	NOAA: *Adapting to Climate Change*
Step 1— Communication and Consultation	Introduction and Step 1: Any stakeholder groups or partners, as well as high-priority concerns	Chapter 5: Build and Maintain Support to Prepare for Climate Change Chapter 6: Build Your Climate Change Preparedness Team	Chapter 3: Establishing the Planning Process
Step 2— Establishing the Context for the Vulnerability Assessment	External resources are not needed for this step	External resources are not needed for this step	External resources are not needed for this step
Step 3— Risk Identification	Step 2: Build a Hazards Profile • 2a: Identify and Examine Current and Past Hazard Risks Step 3: Build a Social Profile • 3a: Identify Local Societal Vulnerability Concerns Step 4: Build an Infrastructure Profile • 4a: Identify Local Infrastructure Vulnerability Factors and Conditions Step 5: Build an Ecosystem Profile • 5a: Identify Key Natural Resources and Conditions	Chapter 4: Scope the Climate Change Impacts to Your Major Sectors	Chapter 4: Vulnerability Assessment • Step 2.1: Identify Climate Change Phenomena • Step 2.2: Identify Climate Change Impacts and Consequences

This Workbook	NOAA: *Roadmap for Adapting to Coastal Risk*	ICLEI: *Preparing for Climate Change*	NOAA: *Adapting to Climate Change*
Step 4— Risk Analysis	Step 2: Build a Hazards Profile • 2b: Explore Climate Trends and Issues Step 3: Build a Social Profile • 3b: Identify Trends and Future Conditions Affecting Societal Vulnerability Step 4: Build an Infrastructure Profile • 4b: Identify Trends and Issues Affecting Future Infrastructure Vulnerability Step 5: Build an Ecosystem Profile • 5b: Identify Ecosystem Resource Stressors • 5c: Identify Trends and Issues Affecting the Future Health of Key Natural Resources	Chapter 8: Conduct a Climate Change Vulnerability Assessment Chapter 9: Conduct a Climate Change Risk Assessment	Chapter 4: Vulnerability Assessment • Step 2.3: Assess Physical Characteristics and Exposure • Step 2.4: Consider Adaptive Capacity • Step 2.5: Develop Scenarios and Simulate Change
Step 5— Risk Evaluation: Comparing Risks		Chapter 9: Conduct a Climate Change Risk Assessment	Chapter 4: Vulnerability Assessment • Step 2.6: Summarize Vulnerability and Identify Focus Areas
Step 6— Establishing the Context for the Action Plan		Chapter 6: Build Your Climate Change Preparedness Team	Chapter 3: Establishing the Planning Process
Step 7— Risk Evaluation: Deciding on a Course		Chapter 4.2: What Will Your Level of Commitment Be?	Chapter 4: Vulnerability Assessment • Step 2.6: Summarize Vulnerability and Identify Focus Areas
Step 8a— Finding Adaptation Actions		Chapter 10.3: Identify Potential Preparedness Actions	Chapter 5: Adaptation Strategy • Step 3.2: Identify Actions

This WORKBOOK	NOAA: *Roadmap for Adapting to Coastal Risk*	ICLEI: *Preparing for Climate Change*	NOAA: *Adapting to Climate Change*
Step 8b—Selecting Adaptation Actions	Step 6: Identifying Strategic Actions	Chapter 9.2: Establish Your List of Priority Planning Areas Chapter 10.4: Select and Prioritize Preparedness Actions	Chapter 5: Adaptation Strategy • Step 3.3: Evaluate, Select, and Prioritize Actions
Step 9—Preparing and Implementing an Action Plan		Chapter 11: Implement Your Preparedness Plan	Chapter 6: Plan Implementation and Maintenance
Step 10—Monitoring and Review		Chapter 12: Measure Your Progress and Update Your Plan	Chapter 6: Plan Implementation and Maintenance • Step 4.4: Track, Evaluate and Communicate Plan Progress

This WORKBOOK is designed to produce a planning-level watershed-scale adaptation plan. Listed below are EPA resources to help evaluate the vulnerability of specific species, sites or sectors. These references may be useful for some especially difficult parts of your vulnerability assessment or when writing project plans in Step 9.

Control Point and Nonpoint Sources of Pollution and Clean Up Pollution

A Screening Assessment of the Potential Impacts of Climate Change on Combined Sewer Overflow (CSO) Mitigation in the Great Lakes and New England Regions. This report is a screening-level assessment of the potential implications of climate change on CSO mitigation in the Great Lakes and New England regions.
http://cfpub.epa.gov/ncea/global/recordisplay.cfm?deid=188306

Maintain and Improve Estuarine Habitat

Climate Change and Interacting Stressors: Implications for Coral Reef Management in American Samoa. The purpose of this report is to provide coral reef managers with management options to enhance the capacity of coral reefs to resist the effects of climate change.
http://cfpub.epa.gov/ncea/global/recordisplay.cfm?deid=173312

Vulnerability Assessments in Support of the Climate Ready Estuaries Program: A Novel Approach Using Expert Judgment. As part of the Climate Ready Estuaries program, EPA's Office of Research and Development has prepared reports exploring a new methodology for climate change vulnerability assessments using San Francisco Bay's and Massachusetts Bays' salt marsh and mudflat ecosystems.

> *Volume I: Results for the San Francisco Estuary Partnership:*
> http://cfpub.epa.gov/ncea/global/recordisplay.cfm?deid=241556

> *Volume II: Results for the Massachusetts Bays Program:*
> http://cfpub.epa.gov/ncea/global/recordisplay.cfm?deid=241555

Aquatic Ecosystems, Water Quality, and Global Change: Challenges of Conducting Multi-Stressor Vulnerability Assessments. This report explores the conceptual and practical challenges associated with using environmental indicators to assess how the resilience of ecosystems and human systems may vary as a function of existing stresses and maladaptations.
http://cfpub.epa.gov/ncea/global/recordisplay.cfm?deid=231508

An Assessment of Decision-Making Processes: Evaluation of Where Land Protection Planning Can Incorporate Climate Change Information. The goal of this report is to assess the feasibility of incorporating climate change impacts into the evaluation of land protection programs. The report concludes that land protection may become more important for jurisdictions, particularly to ameliorate climate change impacts on watersheds and wildlife.
http://cfpub.epa.gov/ncea/global/recordisplay.cfm?deid=238091

Climate and Land Use Change Effects on Ecological Resources in Three Watersheds: A Synthesis Report (External Review Draft). This report provides a summary of the scientific findings from the three case studies and discusses insights gained from a comparison across case studies of the process of conducting watershed assessments and effective ways of improving our capability to support decisions.
http://cfpub.epa.gov/ncea/global/recordisplay.cfm?deid=180083

Integrated Climate and Land Use Scenarios (ICLUS). The ICLUS tools for ArcGIS will allow users to customize housing density patterns by altering household size and travel time assumptions; reclassify housing density into classes different than those already provided; and generate a map of estimated impervious surface based on a housing density map.
http://www.epa.gov/ncea/global/iclus/

Protect and Propagate Fish, Shellfish and Wildlife, Including Control of Nonnative Species

Climate Change and Aquatic Invasive Species. This report reviews available literature on climate-change effects on aquatic invasive species and examines state-level aquatic invasive species management activities.
http://cfpub.epa.gov/ncea/global/recordisplay.cfm?deid=188305

Climate Change Effects on Stream and River Biological Indicators: A Preliminary Analysis. This report is a preliminary assessment that describes how biological indicators are likely to respond to climate change, how well current sampling schemes may detect climate-driven changes, and how likely it is that these sampling schemes will continue to detect impairment.
http://cfpub.epa.gov/ncea/global/recordisplay.cfm?deid=190304

A Framework for Categorizing the Relative Vulnerability of Threatened and Endangered Species to Climate Change (External Review Draft). Modules in this report walk the user through a systematic process for (1) categorizing a species' baseline vulnerability to extinction or major population reduction, (2) categorizing a species' vulnerability to future climate change, (3) developing a matrix that gives an overall score of the species' vulnerability to non-climate and climate change stressors, and (4) qualitatively determining the uncertainty in the estimate of a species' vulnerability.
http://cfpub.epa.gov/ncea/global/recordisplay.cfm?deid=203743

Collaborative Guide: A Reef Manager's Guide to Coral Bleaching. The guide provides coral reef managers with the latest scientific information on the causes of coral bleaching and new management strategies for responding to this significant threat to coral reef ecosystems.
http://cfpub.epa.gov/ncea/global/recordisplay.cfm?deid=159849

Protect Public Water Supplies and Recreational Activities, in and on the Water

Climate Change Vulnerability Assessments: Four Case Studies of Water Utility Practices. The case studies illustrate different approaches that reflect specific local needs and conditions, existing vulnerabilities, local partnerships, and available information about climate change.
http://cfpub.epa.gov/ncea/global/recordisplay.cfm?deid=233808

Climate Ready Water Utilities Climate Resilience Evaluation and Awareness Tool. This software tool assists drinking water and wastewater utility owners and operators in understanding potential climate change impacts and in assessing the related risks at their utilities.
http://water.epa.gov/infrastructure/watersecurity/climate/

Three conceptual model diagrams from EPA's Causal Analysis/Diagnosis Decision Information System are shown below. They might be useful for the target audience of the WORKBOOK. More conceptual model resources are available at the CADDIS website, http://www.epa.gov/caddis/index.html.

The CADDIS website (http://www.epa.gov/caddis/ssr_condiag_popup.html) describes these conceptual diagrams in this way:

"A conceptual diagram is a visual representation of how a system works. In CADDIS, these diagrams are used to describe hypothesized relationships among sources, stressors, and biotic responses within aquatic systems. Conceptual diagrams and accompanying narrative descriptions are useful tools throughout the Stressor Identification process, from structuring initial brainstorming, to providing a framework for data collection and analysis, to organizing and presenting results.

"These diagrams provide overviews of how specific stressors may be linked to sources and biological effects, by illustrating potential linkages among stressors (or candidate causes) and their likely sources and effects based on scientific literature and professional judgment. Inclusion of a linkage indicates that the linkage can occur, not that it always occurs.

...

"The diagrams shown here are meant to serve as starting points—consult these diagrams as you begin to think critically about how these sources and stressors may be operating in your system, and modify them as needed to reflect key components and relationships in your particular stream. For example, you may know that certain sources shown in a diagram are not found in your watershed, but other sources that are not shown are present. Your site-specific conceptual diagram should reflect these differences."

Users should exercise their professional judgment about whether any conceptual model adequately describes their environmental systems.

Temperature: Conceptual Diagram

http://www.epa.gov/caddis/ssr_temp4d.html

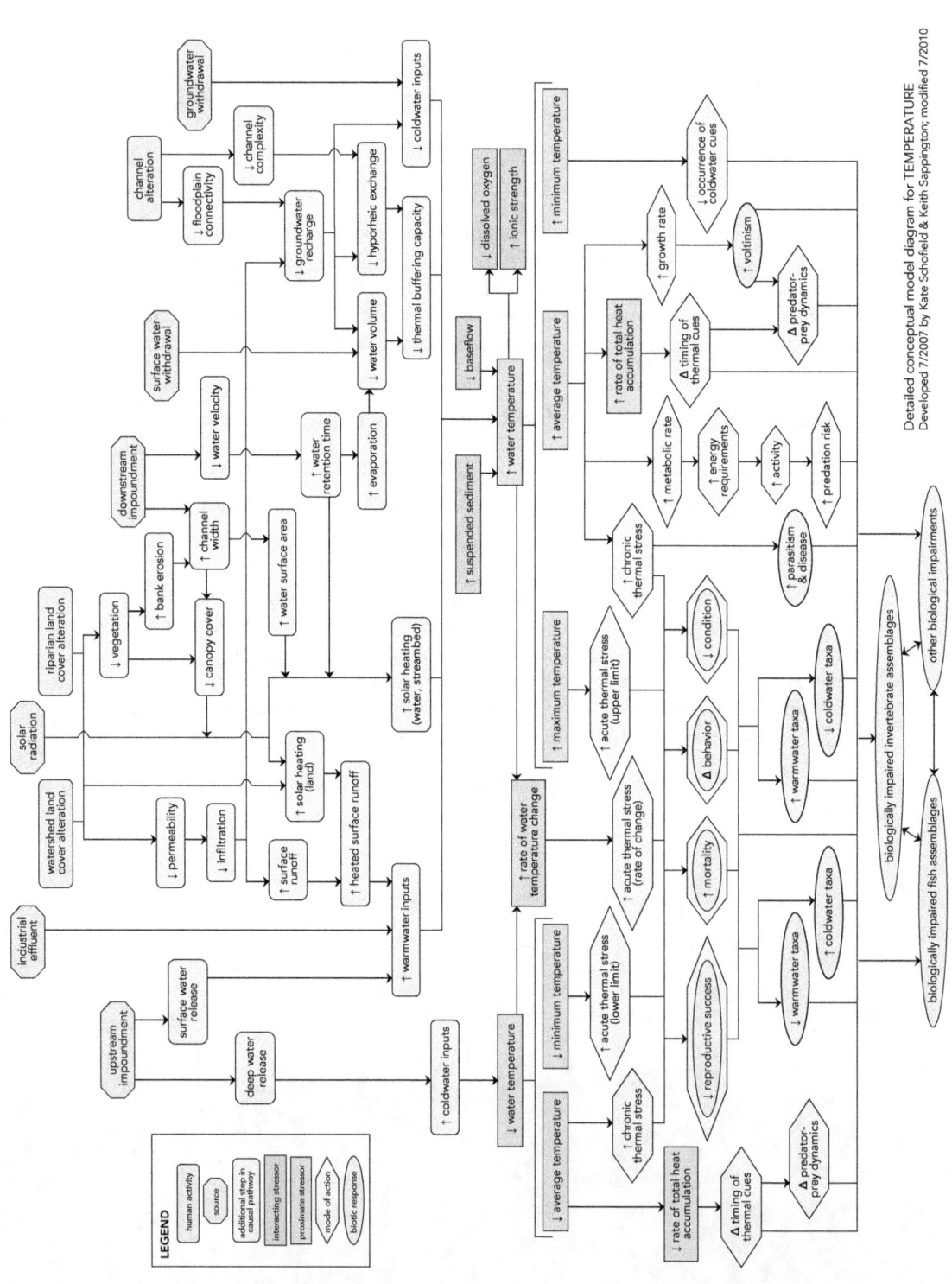

Detailed conceptual model diagram for TEMPERATURE
Developed 7/2007 by Kate Schofield & Keith Sappington; modified 7/2010

Flow Alteration: Conceptual Diagram

http://www.epa.gov/caddis/ssr_flow4d.html

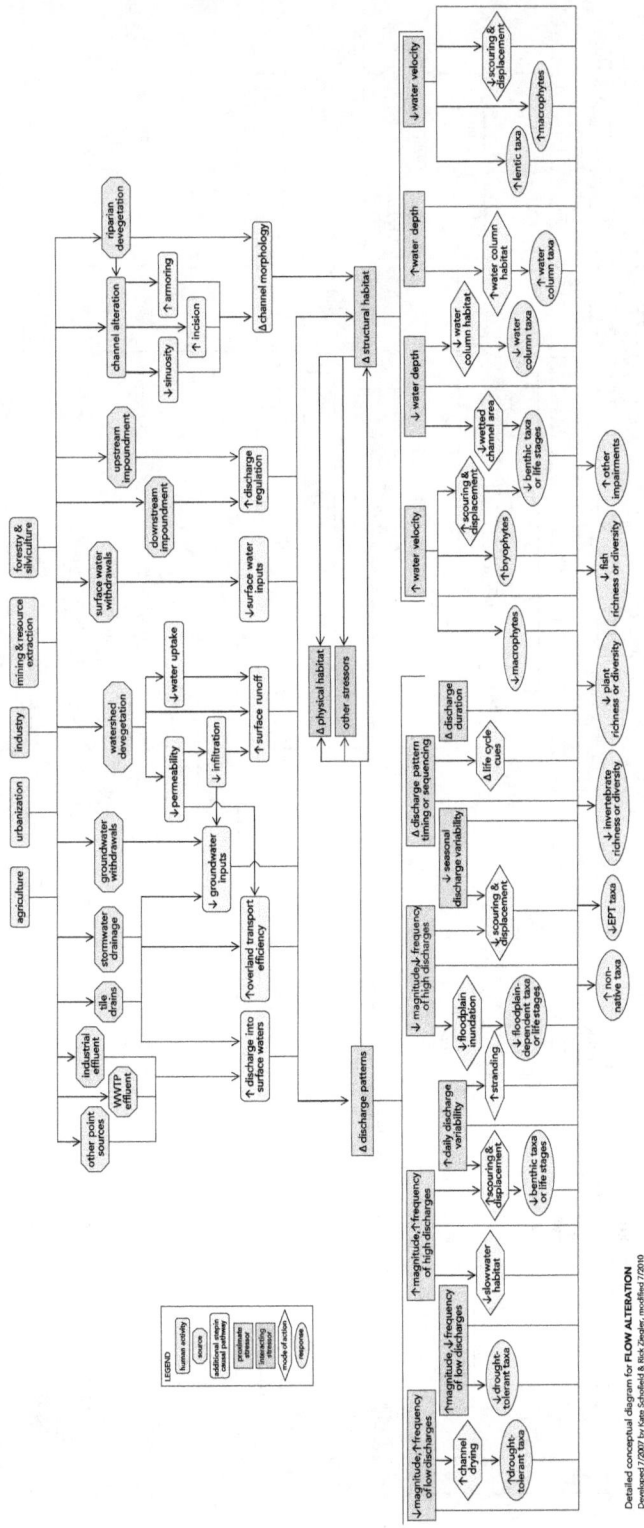

Detailed conceptual diagram for **FLOW ALTERATION**
Developed 7/2007 by Kate Schofield & Rick Ziegler, modified 7/2010

Physical Habitat: Conceptual Diagram

http://www.epa.gov/caddis/ssr_phab4d.html

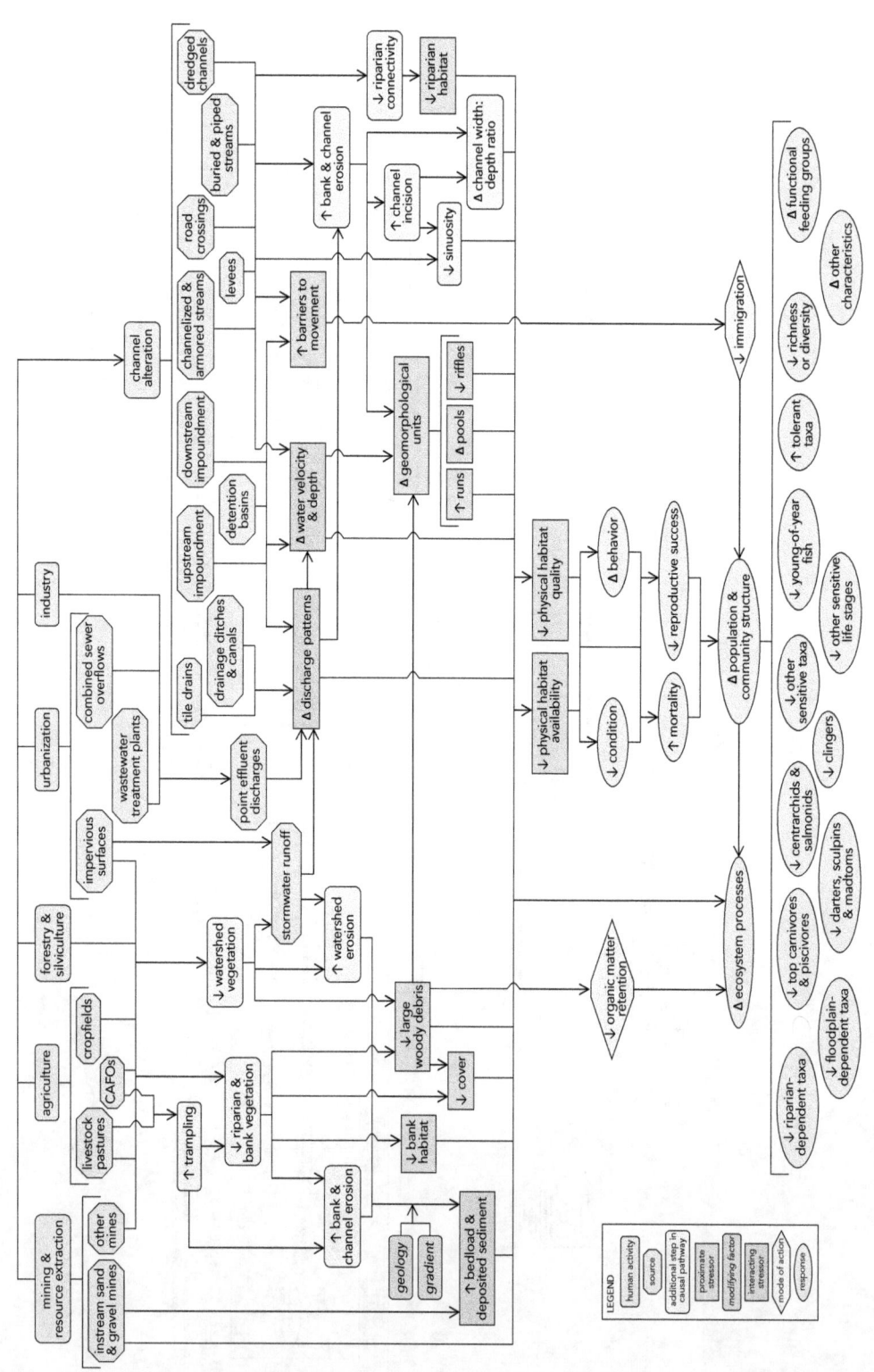

Simple conceptual model diagram for **PHYSICAL HABITAT**

This resource guide was current in May 2014.

U.S. Global Change Research Program

The best single source for climate change projections and climate change impact reports is the U.S. Global Change Research Program.
http://globalchange.gov/

USGCRP periodically issues a National Climate Assessment that informs the nation about already observed changes, the current status of the climate, and anticipated trends for the future. The regional and sector reports could be the primary resource for much of your vulnerability assessment.
http://nca2014.globalchange.gov/

General EPA Resources About Climate Change

EPA: Climate Change
http://www.epa.gov/climatechange

EPA Office of Research and Development: Climate Change Research
http://www.epa.gov/research/climatescience/

EPA: Climate Change and Water
http://water.epa.gov/scitech/climatechange/index.cfm

EPA: Climate Ready Water Utilities Toolkit
http://water.epa.gov/infrastructure/watersecurity/climate/

Climate Change and Coasts

EPA: Climate Ready Estuaries
http://www.epa.gov/cre/

USGCRP. 2009. *Coastal Sensitivity to Sea Level Rise: A Focus on the Mid-Atlantic Region.* Synthesis and Assessment Product 4.1. Part 1: The Physical Environment.
http://downloads.globalchange.gov/sap/sap4-1/sap4-1-final-report-all.pdf

USGCRP. 2008. *Preliminary Review of Adaptation Options for Climate-Sensitive Ecosystems and Resources.* Synthesis and Assessment Product 4.4. Chapter 7: National Estuaries.
http://cfpub.epa.gov/ncea/cfm/recordisplay.cfm?deid=180143

Temperature and Precipitation

The NEX-DCP30 Viewer allows the user to visualize projected climate change for any county in the continental United States.
http://www.usgs.gov/climate_landuse/clu_rd/nex-dcp30.asp

The USGS Derived Downscaled Climate Projection Portal allows visualization and downloading of future climate projections from a group of statistically downscaled global climate models.
http://cida.usgs.gov/climate/derivative/

Drought

Drought is discussed throughout the 2014 National Climate Assessment. Searching the document for "drought" turns up more than 400 results. Searching for "water stress," "dry" or "soil moisture" also leads to useful information about the drought stressor.

Sea Level Rise Viewers

Sea level rise viewers can be a useful tool for visualizing sea level rise impacts (consequence) in your study area. These tools do not, however, indicate the likelihood of any given sea level rise scenario. Use the resources identified above to determine likely scenarios for your area.

NOAA Coastal Services Center: Sea Level Rise and Coastal Hazards Viewer
http://www.csc.noaa.gov/digitalcoast/tools/slrviewer/

The Nature Conservancy: coastal resilience tool (sea level rise visualization for New York and Connecticut)
http://coastalresilience.org/geographies/new-york-and-connecticut

Sarasota Bay National Estuary Program: sea level rise viewer
http://sarasotabay.org/slr-web-map/

Climate Central: Surging Seas sea level rise viewer
http://sealevel.climatecentral.org/

Ocean Acidification from Increased Carbon Dioxide Levels

National Climate Assessment: Oceans and Marine Resources
http://nca2014.globalchange.gov/report/regions/oceans

National Research Council: Ocean Acidification
http://nas-sites.org/oceanacidification/

Rebecca Hansen

If warmer water is determined to be a stressor that leads to unwanted risks, then a number of actions could potentially be effective at mitigating those risks.

TABLE F-1. ACTIONS THAT LOWER WATER TEMPERATURE AND CONTROL URBAN STORMWATER

Action	Water temperature benefits	Other benefits that might come from this action
Planting trees	• Shades the ground and keeps it cooler	• Controls runoff and promotes infiltration
Constructing narrow streets	• Results in less heat-holding asphalt and concrete	• Yields less runoff
Building swales and rain gardens	• Gets water underground and maintains aquifers	• Keeps runoff out of waterways
Using rain barrels and cisterns	• Reduces the need for stream diversions and demand on aquifers for irrigation water	• Keeps stormwater on a lot
Paving with permeable materials	• Keeps runoff from moving over heated roadways and parking lots	• Promotes infiltration where the rain falls
Installing green roofs	• Lowers temperatures compared to conventional roofs; reduces energy use and waste heat	• Traps stormwater on site

TABLE F-2. ACTIONS THAT LOWER WATER TEMPERATURE AND RESTORE WATERSHED STRUCTURE AND FUNCTION

Action	Water temperature benefits	Other benefits that might come from this action
In-stream measures		
Removing unneeded dams and impoundments	• Keeps impounded waters from heating up	• Restores natural hydrology • Returns to natural sediment transport and geomorphology • Reestablishes natural disturbance
Releasing cold water from upstream impoundments	• Strategically lowers water temperature	• Constructs biotic refugia or habitat • Builds biological communities
Controlling stream bank erosion	• Keeps channels from getting wider and shallower and warming more easily	• Returns to natural sediment transport and geomorphology • Raises water quality
Creating deep pools or artificial logjams	• Provides shade or deep water that limits direct heating from sunlight	• Constructs biotic refugia or habitat • Builds biological communities
Groundwater measures		
Controlling groundwater withdrawal	• Maintains groundwater sources that supply base flow to streams	• Creates habitat and hydrologic connectivity • Restores natural hydrology
Promoting stormwater infiltration	• Gets water into aquifers and away from exposure to the sun • Recharges groundwater that supplies baseflow that regulates stream temperature	• Restores natural hydrology • Returns to natural sediment transport and geomorphology • Reestablishes natural disturbance
Removing unneeded channelization	• Restores natural groundwater exchange • Restores connection to floodplains which promotes floodwater infiltration into aquifers	• Restores natural hydrology • Returns to natural sediment transport and geomorphology • Reestablishes natural disturbance
Land use measures		
Planting forest and floodplain habitat	• Shades watershed lands, surface waters and streambeds • Reduces runoff and promotes groundwater infiltration	• Creates habitat and hydrologic connectivity • Rebuilds native vegetation and corridor networks • Raises water quality
Keeping livestock out of streams	• Reduces bank erosion	• Returns to natural sediment transport and geomorphology • Raises water quality
Controlling soil erosion in the watershed	• Keeps sediment from clogging streambeds and interfering with groundwater exchange • Keeps heat-trapping particles out of waterways	• Returns to natural sediment transport and geomorphology • Raises water quality
Controlling stormwater runoff	• Reduces high peak flows that contribute to erosion and channel changes	• Restores natural hydrology • Returns to natural sediment transport and geomorphology • Reestablishes natural disturbance • Raises water quality

References

For more information please see the following resources:

Impairments from warmer water

EPA. 2012. Sources, Stressors & Responses—Temperature. CADDIS Volume 2.
http://www.epa.gov/caddis/ssr_temp_int.html

EPA Region 10 water temperature guidance:

- EPA. 2003. *EPA Region 10 Guidance for Pacific Northwest State and Tribal Temperature Water Quality Standards*. EPA 910-B-03-002.
  http://www.epa.gov/region10/pdf/water/final_temperature_guidance_2003.pdf
- EPA. 2001. *Issue Paper 3: Spatial and Temporal Patterns of Stream Temperature (Revised)*. EPA 910-D-01-003.
  http://yosemite.epa.gov/R10/WATER.NSF/6cb1a1df2c49e4968825688200712cb7/5eb9e547ee9e111f88256a03005bd665/$FILE/Issue%203%20Spatial%20Temp.pdf

USGS Water Science School. 2014. Water Properties: Temperature.
http://ga.water.usgs.gov/edu/temperature.html

Managing unwanted warming

EPA. 2008. *Reducing Urban Heat Islands: Compendium of Strategies*.
http://www.epa.gov/heatisld/resources/compendium.htm

EPA. 2014. Green Infrastructure.
http://water.epa.gov/infrastructure/greeninfrastructure/index.cfm

EPA. 2014. Healthy Watersheds.
http://water.epa.gov/polwaste/nps/watershed/index.cfm

EPA. 2010. *A Method to Assess Climate-Relevant Decisions: Application in the Chesapeake Bay*. External Review Draft. EPA 600-R-10-096A.
http://cfpub.epa.gov/ncea/cfm/recordisplay.cfm?deid=227483

University of New Hampshire Stormwater Center. 2011. *Examination of Thermal Impacts from Stormwater Best Management Practices*.
http://www.unh.edu/unhsc/sites/unh.edu.unhsc/files/progress_reports/UNHSC%20EPA_Thermal_Study_Final_Report_1-28-11.pdf

T. Beechie, H. Imaki, J. Greene, A. Wade, H. Wu, G. Pess, P. Roni, J. Kimball, J. Stanford, P. Kiffney and N. Mantua. 2012. Restoring Salmon Habitat for a Changing Climate. *River Research and Applications*. DOI: 10.1002/rra.2590

Website links verified in July 2014

www.ingramcontent.com/pod-product-compliance
Lightning Source LLC
Chambersburg PA
CBHW080641180526
45168CB00008B/3263